U0271357

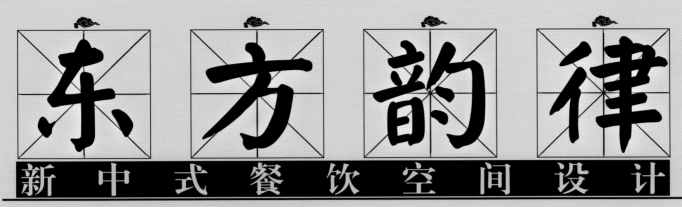

东方韵律
新 中 式 餐 饮 空 间 设 计

LINGERING CHARM OF THE ORIENT 精品文化 编
NEW SPACE DESIGN OF CHINESE RESTAURANTS

华中科技大学出版社
http://www.hustp.com
中国·武汉

图书在版编目(CIP)数据

东方韵律:新中式餐饮空间设计/ 精品文化编.—武汉:华中科技大学出版社,2016.1
ISBN 978-7-5680-1010-8

Ⅰ. ①东… Ⅱ. ①精… Ⅲ. ①餐馆—室内装饰设—图集 Ⅳ.①TU247-64

中国版本图书馆CIP数据核字(2015)第148198号

东方韵律　新中式餐饮空间设计

精品文化　编

出版发行:华中科技大学出版社(中国·武汉)
地　　址:武汉市武昌珞喻路1037号(邮编:430074)
出 版 人:阮海洪

责任编辑:曾　晟　　　　　　　　　　　　　　　　　　　　　　　责任监印:秦　英
责任校对:刘锐桢　　　　　　　　　　　　　　　　　　　　　　　装帧设计:李　华

印　　刷:深圳当纳利印刷有限公司
开　　本:969 mm×1270 mm　1/16
印　　张:20
字　　数:288千字
版　　次:2016年1月第1版　第1次印刷
定　　价:328.00元(USD 51.69)

投稿热线:(010)64155588-8000
本书若有印装质量问题,请向出版社营销中心调换
全国免费服务热线:400-6679-118　竭诚为您服务
版权所有　侵权必究

序言 PREFACE

For some people, design is interior decoration, is to select curtain fabric sofa. But for me, nothing is more important than design philosophy. Compared to restaurant food, the unique interior design style and its ethos revealed are more appealing to me. It does not require much complex and cumbersome decoration, but always touched by the exquisite detail in the accidental explore. When you walk into a restaurant, the overall tone of chilly color spreading space pervade out a peaceful atmosphere, under the decoration of nostalgic chandelier, low-key backdrop in light gray covered with a touch of art, and simple meal tables and chairs with a nostalgic and dependence on the home, making you review the dream of the old days.

And I personally feel that when design restaurant space; we should pay attention to the functional requirements of practitioners and the implicit demand of customers. Also we need a good lighting design, because good lighting design can soften the hard environment of space, making people feel comfortable, the lights on the table should bring out the food looks, so that the original delicious food looks more attractive visually. And candlelight placed on or around the table enhances the romantic atmosphere of entire space, which will make people remember the unique experience deeply brought by the design while in the process of enjoying food.

The color of environment is the soul in environmental design, which also has a significant impact on the sense of the space, comfort, atmosphere of interior design and the human psychology and physiology. Within a fixed environment, the first thing burst into our visual senses is the color, and the most contagious is also color. It produces beautiful reverie from pleasing harmony color, making environment into feelings. Now the promotion of human and personalization has become the mainstream of modern environmental design, only put people first, can make the design humanize and narrow the distance between human and human, only make deep research in different people can create a personalized environment design, and only adhere to the "people-oriented" purposes, can create out the ideal design environment. The spirit of the interior design is to influence people's emotions, and even affect will and action, only deeply understanding of people's living habits, culture, behavior psychology and visual perception that we can create a good design.

Design is everywhere in life, design comes from life but more than life, they are interdependent and mutually reinforcing. Nature, life provides fertile soil to design, and the design performance nature and life to the fullest, making life more colorful. Life comes from nature, Life is full of diversity, things, nature is endless variation, and therefore, the design is colorful.

Fujian host architectural design company Li ChuanDao

设计对于有些人而言，它是室内装饰，它是窗帘沙发的布料选择。但对于我而言，没有能比设计思想更重要的东西了。一个餐厅相较于美食更吸引我的是它内部独特的设计风格和它透出的精神气质。它不需要太多复杂繁琐的装饰，只需让人在偶然的探寻中被精致的细节所触动。当你走进一家餐厅时，清冽的色彩铺开了空间的整体基调，弥漫出安然、静雅的氛围，低调的浅灰色的背景墙在怀旧吊灯的装饰下，带着淡淡的艺术气息，而质朴的餐桌椅带着一份对家的眷恋和依赖，让人重温旧时光的梦想。我觉得餐饮空间的设计需要注重的是餐饮从业者的功能性需求，以及客户对空间所感受的隐性需求。它还需要一个好的灯光设计，因为好的灯光设计能让坚硬的环境空间造型柔化，让人感觉舒服，餐桌上的灯光衬托出食物的外观，让原本可口的美食在视觉上更加诱人。而放在餐桌及四周的烛光提升了整个空间浪漫、感性的气氛，这一定会让人在享受美食的过程也深深记住了设计给人带来的独特体验。环境的色彩是环境设计的灵魂，环境色彩对室内外设计的空间感、舒适度、环境气氛及对人的心理和生理都有很大的影响。在一个固定的环境中最先闯入我们视觉感官的是色彩，而最具有感染力的也是色彩。人们可从和谐悦目的色彩中产生美的遐想，化景为情。而今提倡人性化和个性化已经成为现代环境设计的主流，只有把人放在第一位，才能使设计人性化，拉近人与人之间的距离。只有对不同的人做深入的研究，才能创造出个性化的环境设计，也只有坚持"以人为本"的宗旨，才能营造出理想的环境。而室内设计的精神就是要影响人们的情感，乃至影响人们的意志和行动，只有深入了解人们的生活习惯、文化内涵、行为心理和视觉感受，才能创造出好的设计。生活中设计无处不在，设计来源于生活，而高于生活，它们相互依存，相互促进。自然、生活给设计提供了肥沃的土壤，而设计尽情地表现自然、表现生活，并使生活更加丰富多彩。生活来源自然，自然是变化无穷的，生活是千姿百态的，而设计是丰富多彩的。

福建东道建筑装饰设计有限公司　李川道

006	Smiling Buddha 微笑浮屠		088	Lushe Dining Club 麓舍餐饮会所
018	Design Concept for Beverley Meizhou Dongpo Restaurant, Los Angeles, USA 美国眉州东坡洛杉矶比佛利店		096	Hakka Impression Traceability Club 印象客家溯源会所
024	Meizhou Dongpo Restaurant, Vanke Midtown, Suzhou 眉州东坡苏州万科美好广场店		104	Lian Yugang Seafood Restaurant 连渔港海鲜餐吧
034	Lanyue Seafood Buffet Restaurant of Fuzhou 福州澜悦海鲜自助餐厅		110	Jiangxi Hengmao Resort Hotel's Chinese Restaurant 江西恒茂度假酒店中餐厅
042	Hai Tiange Chinese Restaurant of Jiali Central Hotel 嘉里中心酒店海天阁中餐厅		120	Nanxincang Branch of Dadong Chain Restaurant 大董餐厅南新仓店
048	First-rate Delicious 原汤的记忆		126	DOZO Izakaka Dining Bar DOZO 创作料理
056	Co-exist & Harmonious 共生·和谐		138	Huludao Food House Private Dining 葫芦岛食屋私人餐厅
064	KARUISAWA Tainan Store 轻井泽台南店		146	Yiran Ju 怡然居
072	Yonghe Club No. 33 雍和会 33 号		156	Wong's Hot Pot Restaurant 王家渡火锅餐厅
080	Elegant Leisure 寻幽竹篁里		164	Yiyun Xiang Restaurant 溢云香餐厅
			172	Seasons Blessing of Duck Restaurant, Mongkok 四季民福烤鸭店旺角店

180 The Taste of Chao Shan 潮汕味道	254 Chating Restaurant Qiaoting Fish Town 桥亭活鱼小镇茶亭店
186 Donglaishun Tianyuan Restaurant 东来顺天元店	260 Hui Jiangnan Private Kitchen 汇江南私房菜
196 Pepper Meets Chili 花椒遇见辣椒	268 Rongfu Restaurant 蓉府餐厅
204 Tori Talk 东篱·叙	276 Sanshili Alley Restaurant 三市里胡同餐厅
212 Maolu Impression Weifang Kaide Restaurant 茅庐印象潍坊凯德店	284 Shicai Yunnan Cuisine Restaurant 食彩云南料理餐厅
220 Maolu Impression Weifang Wanda Restaurant 茅庐印象潍坊万达店	290 Qian Yuan Fashion of Tianjin 天津乾园风尚
226 Mashijiu Pot Aksu Restaurant 马仕玖煲阿克苏店	296 Taste Talk Restaurant 味语餐厅
234 Mashijiu Pot Changchun Road Restaurant 马仕玖煲长春路店	302 Story of Little Tow Retro Romantic Feelings 小城故事复古情怀
240 South Pavilion Restaurant 南堂馆餐厅	310 Number.1 Zhuxi 竹溪一号
248 Spicy Talking Sichuan Restaurant Daming Outlet 说麻道辣川菜馆达明店	316 The Mutton Soup of Eight Banners 八旗羊汤

CONTENTS

地点	面积	设计师	设计公司
/台湾	/1500平方米	/周易	/周易设计工作室

Smiling Buddha
微笑浮屠

The meaning of commercial restaurant to modern people, in addition to the satisfaction to taste buds, is it possible to have another level? For example, the extension or description on the spiritual aspect? Of course! Overthrowing the restaurant model which people generally known, also examining the market acceptance, JOY interior Design Studio uses clean vocabulary such as Buddha hand, Buddha head, floating candles and slender stick of incense in blending the space atmosphere, building a theme space which closely combines delicious hot-pot cuisine and fantasy creation, through the leisurely rhyme of figurative and abstract interpretation to iron the emotional ups and downs of human beings.

Clean Mind

Gray building located on the corner land is stable and rustic like a castle, mottled background color and the windows on the lower limit of both sides, restrainedly passed a concept similar to a private club, the laminating iron shell words are inlayed on the front face, on the place, the three words "Tien Shui Yueh" were written elegantly, making the brand awareness at once, it is also like the morning bell, knocking directly into the heart of the overs. The external arcade and cloister introduce the literati romantic appearance of Suzhou garden, the delicate man-shaped puzzle on the floor guide visitors' footstep, a line of waiting chairs made by inoculating old railroad crossties and iron foot is echoing the slender column lights which supporting the rain cover roof, on the waterscape floating islands between corridors and main buildings, bamboo swaying green and mountain fern are carefully planted showing the intoxicating poetry like the "bamboo-leave drip with a music of dew" described by a famous poet, Meng Haoran.

Towering up

Push open the wooden doors carved with cloud pattern, the huge gilt Buddha hand towering up the sky on the both sides, the bracketing up gesture is astonishing, as if it is about to push open the black-painted roof, it is also the most attractive sights for checking into places. Straight ahead along the line of sight, up to seven meters three dimension sculpture of Buddha head with closing eyes and smile stands hanging at the end, it contrasts distinct levels under the scoring of the bottom lights, the chin of Buddha head is just hanging above the water mist, the curve of the upward smiling lip exudes the charm of love making restless people feel more secure and eased. Especially, a long rectangular mirror-like waterscape links the Buddha head and Buddha hand, the projected light embedded in the both sides of the base built with grayscale sip stone bricks and the two lines of hand-made glass candle lights jointly construct a fantasy light, and there is always dense water mist circling in the middle which contrasts finely with the incense devices hanging from the vertex of the roof.

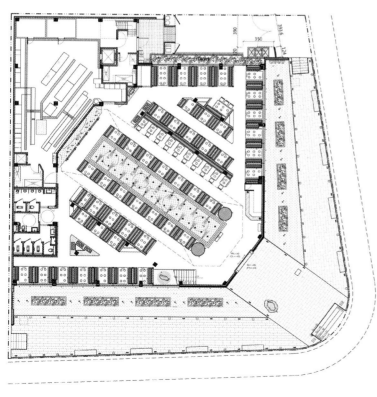
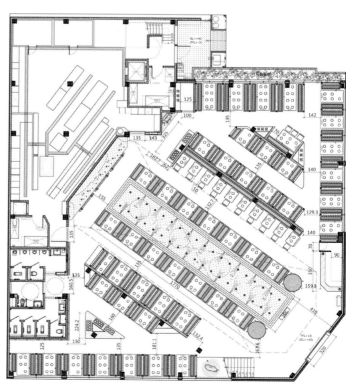

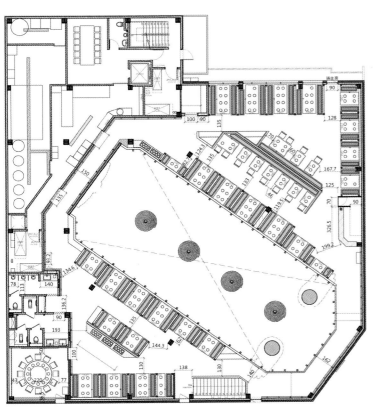

Drum music
Forming the details of dramatic contrast with the quaint and ethereal space is the incidental music layout that the designer using movie scene thinking. Unlike the faintly discernible traditional stringed and woodwind instruments music, it is the drumming with clear and strong sense of rhythm, in the warm and majestic melody, there is passionate appeal in the vague ritual, and such kind of attention and creativity with multi-sensory powers dramatically strengthen the concreteness of the proposals.

Fertility
Dining area on the first floor and the central waterscape are showing a parallel row-line crossing pattern, the design of seat hardware was deliberately using black color style to weaken the hardware and making it as a part of the background, iron-made fine grille is used to define and separate the seats, maintain beauty of the penetration of perspective and tranquility, project the lights to the bottom of the seats to increase the light emitter layers. JOY design studio has always been good at lighting design and it has vividly showing in the design of "Tien Shui Yueh", all scene lights, lighting modeling, and illumination color temperature all have been planned up in details in advance to ensure the precise focus without any interference, the ever-changing visual feast is the best condiments of cuisine.

Submergence
High ceiling space is also another big advantage in this case, elevating along the line of sight up, the old wood stacking along the two sides of the wall, interpreting the old but warm feeling of time, wood texture under the thin wave of lights showing a curving type of rough. The corridors on the second floor connecting the dining areas on the both sides are covered with plenty of wood cladding profiles, the level difference between the shape of the surface highlighting the simplicity and natural aroma of wood, ground lines and wall lamps moderately complete the moving line guide in the fantasy space. The fertility of spiritual aspects and the momentum generated by multi-elements clustering making the unique situation atmosphere in "Tien Shui Yueh" every-lastingly immersed in the crystalline state after the concentration of ancient oriental aesthetic enrichment.

商业餐厅对于现代人来讲，除了味觉上的满足，还有没有其他层次的意义，例如精神层面的延伸或阐述？这当然是有的，本案颠覆一般人印象中的餐厅形象，也考验市场的接受度。设计以佛手、佛头、浮烛和线香等清净之物调和空间氛围，打造了一处将美味食物与奇幻风格相契合的主题空间。

净观
建筑宛如城郭般安定质朴，斑驳底色加上两侧低限开窗，内敛地呈现着类似私人会所的氛围。在正面嵌上发光的铁壳字，上面写着飘逸的"天水玥"三字，像是清晨的梵钟一般，直直敲进观者的心坎。外观骑楼回廊引入苏州园林的人文浪漫，地面精致的铺设引导步履，古旧枕木与铁制作而成的等待椅一字排开，与修长柱灯相呼应，回廊和主建筑之间的水景浮岛上，精心种植了随风摇曳的翠绿幽竹和山蕨，颇有孟浩然笔下"竹露滴清响"的醉人诗意。

擎天
推开镂刻云纹的木雕大门，两侧巨大的描金佛手擎天而立，向上托撑的手势相当摄人，仿佛要推开屋顶，这也是最吸引来客的景点。沿此处直行，视线向前，尽头处高达七米的立体佛头雕塑垂目浅笑，在底部光源烘托下层次分明，佛首的下巴处恰好悬浮于迷离水雾之上，唇角上扬的弧线散发出慈爱庄严的神韵，让躁动的人心倍感安稳。特别是佛头与佛手之间，以一座长矩形镜面水景串联，石子砌成的基座两侧内嵌投射灯光，与悬浮于水面上的两列手工玻璃烛灯共构梦幻光影，水雾腾绕其间，与从顶棚垂挂而下的装置相映成趣。

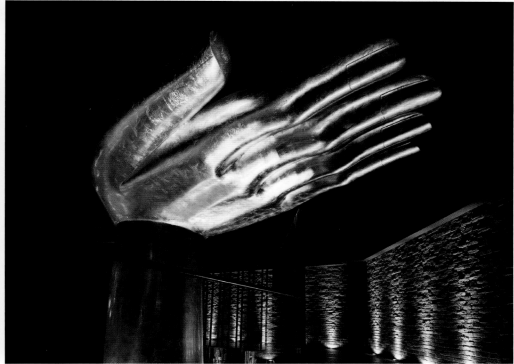

鼓乐
和空间的古朴气质形成巨大反差的细节，应该是设计师以电影场景思维布局的情境配乐。不同于若有若无的丝竹之乐，设计师采用节奏感明确、强烈的鼓乐，热情、磅礴的旋律里，隐约有种祭典仪式情绪激昂的感染力。

丰饶
一楼用餐区与中央水景呈平行的行列格局，在座椅的设计上设计师刻意以黑色调弱化处理，使之成为背景的一部分，单一卡座间以铁制细格栅界定，维持视角的穿透感与宁静之美，座椅底部投光以增加光影层次。设计师一贯擅长的灯光设计，在"天水玥"里彰显的淋漓尽致，现场所有的情境光源、灯饰造型、照度色温，都经过事前详尽的沙盘推演，确保精准地烘托目标重点而互不干扰，千变万化的视觉飨宴成为美食佳肴的最佳佐料。

沉潜
空间的挑高也是本案一大优势，随着仰角视线往上，可看见两侧墙面以老木排列，诠释了老旧但温暖的时光，木头的肌理在灯光下别有一番刀劈斧凿的粗犷。二楼衔接两侧用餐区的回廊以大量原木剖面贴覆，造型面的高低差，彰显出木头天生的纯朴与香气。地面的线条与墙面的壁灯，适度在空间中完成动线引导。精神层面的丰饶加上多种风情元素齐聚的气势，让本案具有与众不同的情境氛围，使顾客沉浸在古老的东方美学中。

地点	面积	设计师	设计公司	主要材料
/洛杉矶	/480平方米	/王砚晨、李向宁	/经典国际设计机构（亚洲）有限公司	/耐候钢板、黄铜板、印刷玻璃、中国黑石材、仿汝窑瓷器

Design Concept for Beverley Meizhou Dongpo Restaurant, Los Angeles, USA

美国眉州东坡洛杉矶比佛利店

Meizhou Dongpo Restaurant devotes its efforts to inherit and carry forward the dongpo food culture. As the first Meizhou Dongpo Restaurant in the United States, its design theme must be thick and heavy in colors so as to express the culture essence. Due to space limitation, we can only select the most representative items for Mr. Su Dongpo's spirits, and arrange them all over the inner and outer space of the restaurant in the United States. These include 1000 Chinese writing brushes, 500 pieces of Ruyao rare variety of chinaware of the Song Dynasty, Mr. Su Dongpo's most representative poems (The First and Second Odes on Chibi, Meditating on the Past at Chibi), the artistic calligraphy and painting work by Mr. Su Dongpo, and Mr. Su Dongpo's favorite-bamboo. Re-arranging and innovating all these items carrying a variety of cultures from Mr. Su Dongpo and the great age, and adopting delicate, exquisite and simple Song style furniture, decorative lighting and ornaments in the space, the space image of the first Meizhou Dongpo Restaurant in the United States is thus collectively established, conveying a modern charm from the broad and deep Dongpo feelings with 1000 years history.

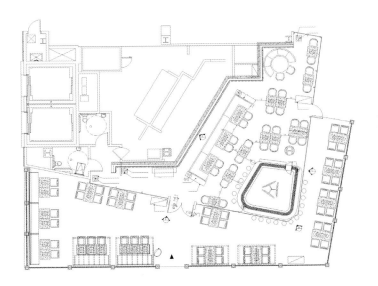

眉州东坡致力于传承及弘扬东坡美食文化。作为眉州东坡登陆美国的第一间餐厅，设计的主题必然是以浓墨重彩的手法来表达东坡文化的精粹。由于空间的体量限制，设计师只是精选最具东坡先生的精神象征的物品，全面呈现于美国店的室内外空间中。其中有1000支中国毛笔、500件宋代汝窑瓷器，最能代表东坡先生的两赋（前、后《赤壁赋》）一词《念奴娇·赤壁怀古》、东坡先生精妙的书法及绘画，还有东坡先生的最爱——竹子。这些与东坡先生及那个伟大时代紧密联系的种种文化载体，经过重组和创新，与空间中精致、细腻、简约的宋风家具、灯具、饰品一起共同组成了眉州东坡美国第一家店的空间物象，传递出一千年来博大而深厚的东坡情怀的时代魅力。

地点	面积	设计师	设计公司	主要材料
/江苏苏州	/1666平方米	/王砚晨、李向宁	/经典国际设计机构（亚洲）有限公司	/加绢玻璃、手工青砖、青铜屏风

Meizhou Dongpo Restaurant, Vanke Midtown, Suzhou

眉州东坡苏州万科美好广场店

Suzhou is known as a paradise on earth, and Suzhou Garden is the best example of the paradise on earth.

When visiting Suzhou Garden, the largest attraction is the application of so called borrowed scenes and symmetry places in the Chinese garden. Exquisite Chinese gardens stress "walking King vary". Moreover, the Chinese literati building gardens tried to represent the space and structure of the outside world perfectly in a limited space. In the Gardens, buildings and pavilions are connected with winding verandas and paths; both inside and outside spaces blend together; through the exquisite and meticulous grilles, vast natural scenery is concentrated into miniature landscapes, taking viewers from tangible reality into the dream of infinite reverie.

In this view, in the designing of Suzhou's first Meizhou Dongpo Restaurant, based on Suzhou's classical gardens, we follows the Chinese literati's gardening concept, adjusting measures to local conditions, borrowing scenes, adopting symmetry places, splitting scenes, separating scenes, and taking a variety of other techniques in organizing spaces. Capturing the essence of Suzhou Garden's visual language – tracery, veranda, folding screen, ebony wooden door and small table etc., we utilize the material of Suzhou features – spun silk, bronze, carved product, handwork blue brick etc., and restructure them with contemporary perspective innovation, collectively creating a winding, representative landscape garden. Being full of poetic, the garden represents both virtual and actual situation, just like traditional Chinese literati's freehand brushwork paintings. The garden has the artistic conception conveyed by the famous little kiosk (its name is "Who can sit with me in this pavilion".) in Humble administrator's garden, being the portraiture of Mr Su Dongpo's mood – "Who can sit with me? Only the moon and breeze".

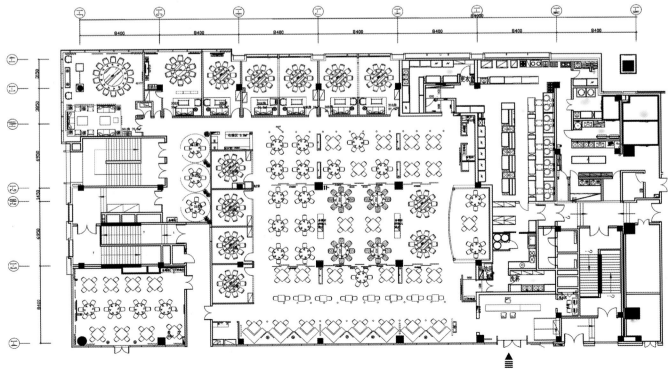

苏州被誉为"人间天堂",苏州园林便是人间天堂的范本。

游苏州园林,最大的看点便是借景与对景在中式园林中的应用。中国园林讲究"步移景异",中国文人造园更是试图在有限的内部空间里完美地再现外部世界的空间和结构。园内亭台楼榭遍布,游廊小径蜿蜒其间,内外空间相互渗透。透过精美细致的窗,广阔的自然风光被浓缩成微型景观,把观赏者从可触摸的真实世界带入无限遐想的梦幻空间。

在眉州东坡苏州首家店的空间设计中,设计师以苏州古典园林为蓝本,遵循中国文人的造园理念,因地制宜,采用借景、对景、分景、隔景等种种手法来组织空间。撷取苏州园林的视觉语言精髓——漏窗、游廊、屏风、檀扇、案几等,运用最具苏州特色的材料工艺——绢丝、青铜、镂刻、手工青砖等,以当代视角创新重组,共同营造出曲折多变、小中见大、虚实相间的充满诗情画意的文人写意山水园林。正如拙政园中那座著名的小亭"与谁同坐轩"所传递出的意境,也是东坡先生"与谁同坐?明月、清风、我"的真实写照。

东方韵律 新中式餐饮空间设计 -029-

地点	面积	设计师	设计公司	主要材料
/福建福州	/2800 平方米	/林昌惠	/研舍设计机构	/橡木、黑木纹大理石、仿铜不锈钢

Lanyue Seafood Buffet Restaurant of Fuzhou

福州澜悦海鲜自助餐厅

This case adopts the new Chinese style to create an oriental style at overall space in general. In the decoration of the hall, a traditional Chinese symmetry technique was used to express the elegant and harmony. The waiting area beside the hall provided a quieter and relax environment by placing some terrine, plots and log tea tables there. Across a special aisle with Chinese style view, the central dining hall was reached, in which both the dining area and the self-help area were constructed with oak log and marble with black wood stripes. The oak stereoscopic grille on the top perfectly combined with the circumspect lighting, which created a kind and elegant circumstance for meal.

On the use of material, the overall space was coated with dark and warm tone by combining retro oak with marble with black wood stripes to avoid being noisy or messy. Moreover, some lively colors were also adopted in local to active the atmosphere. And the copper imitation dealing in details also made the space more elegant and exquisite.

本案的设计风格以新中式风格为主，营造出整体空间的东方气韵。在门厅的装饰中运用了传统中式的对称手法，以体现高雅与和谐，而门厅边的候客区则以更为休闲的处理方式展现，陶罐、盆栽、原木茶桌使客人在此等候时更加静心与放松。通过一条特别设计的中式景观入口走廊便可进入餐厅的中心用餐区。这里不管是用餐区还是自选区，都以橡木原木与黑木纹大理石为主要材料。顶部橡木立体格栅的装饰与灯光设计完美结合，营造出亲切、典雅的用餐环境。在材质应用上，为避免空间环境的嘈杂与零乱，空间整体采取暗色调与暖色调，将复古色橡木与黑木纹大理石材质结合。局部软装使用少量跳色来活跃餐厅氛围，仿铜金属材质的细节处理使空间更显高雅与精致。

地点	面积	设计师	设计公司
/北京	/1087 平方米	/林伟尔	/思联建筑设计有限公司

Hai Tiange Chinese Restaurant of Jiali Central Hotel

嘉里中心酒店海天阁中餐厅

The project is specially designed for contemporary Chinese leisure restaurant.As for materials, the designer applies same materials of wooden screens, lanterns, veneers or stone floors to build the "Zen" atmosphere. At the same time, the wall itself and big art works such as lamplights and wallpaper are decorated for the space, to make the space more elegant and beautiful. The aim of internal space design in the restaurant is to combine openness and privacy as well as build a variery of dining environment. The open-plan kicthen is aslo the focus of primary dining space.The wooden screen design is derived from traditional grid screen, to partition dining space to form semi-private dining area.

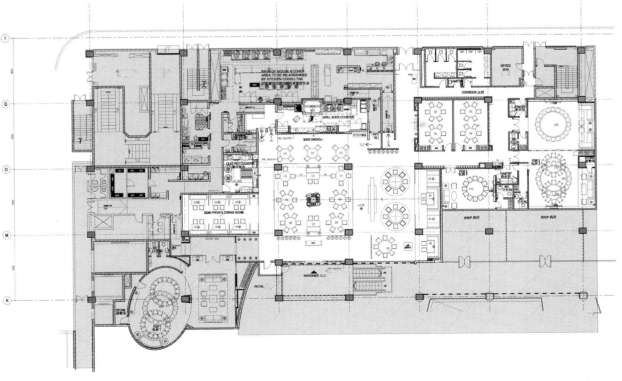

本案是专为中式休闲餐饮而设计的。材料上,设计师使用了材质统一的木质屏风、灯笼、木饰面或石材地板,营造出空间"禅"的氛围。同时,空间中以大量的主题墙身和大型艺术品作为点缀,如灯饰、壁纸等,都使空间更具气质。餐厅室内空间设计的目的是,在结合开放性和私密性的同时创造出各种不同的用餐环境,开放式厨房成为主要用餐空间的焦点。木屏风设计来源于传统格子屏风,以屏风分隔用餐空间,形成半私密餐饮区域。

地点	面积	设计师	设计公司	主要材料
/台湾新竹	/498平方米	/周易	/周易设计工作室	/青砖、杉木、竹、铁刀木皮、抿石子、磨石砖、非洲花梨木、锈铁

First-rate Delicious
原汤的记忆

The project is a detached building with four storeys, not only from distinctive architectural style in appearance, but also from landscape art of high degree of recognition. All of them form a more signifant landmark of the project.

There are the arcade and the staircase circulation on the first floor, but available space is limited, so the designer places a modest reception counter and waterscape, to adorn classical stones and aquatic plants. Decorated with table bricks in direct eyesight, the first floor eases custmers' mood at the first time and leads them to their seats.

Besides exquisite booths are circled by grid behind the first floor, scenery planning above the second floor are more wonderful. The designer places a terraced stone screen by the staircase, and makes water fall and lamplights on both sides which are worthing seeing, to keep a distant vision. The other three floors are all dining space full of fun except for one kitchen on the fourth floor.

The designer applies Japanese stone lamplights, bricks, drifting woods, suspended bamboo pipes, antique doors and other elements into the space, combines these natural materials full of passageway of time with Japanese garden essence, waterscape like a mirror, soft lamplights and even carved words by stained iron on the wall, and brews a harmonious fusion of Chinese and Japanese culture.

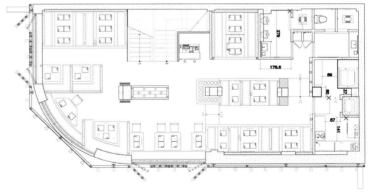

本案是一栋四层楼高的独立建筑，不管是外观上鲜明的建筑形式，还是高辨识度的地景艺术，都让本案更具地标性。

一楼因为包含骑楼与扶梯，可用空间有限，因此设计师在进门处设置了质朴稳重的接待台和流动水景，用古朴的石钵与水生植栽点缀，在青砖立面衬托下，第一时间舒缓来客情绪并引导入座。

相较一楼后方以半腰格栅界定的精致卡座，二楼的用餐区的规划更为精彩。设计师在梯口处设计了石砌梯形屏风，融入两面皆可观赏的流瀑和灯光，维持了视觉上的穿透。除了四楼有备餐厨房外，其他三个楼层皆是富有情调的用餐空间。

设计师在空间中融入日式石灯、青砖、漂流木、悬浮在空中的竹管群、古董大门等元素，整合这些充满怀旧感与时间感的自然材质，加上日式庭园、镜面水景、柔和的灯带照明，甚至是墙面上以锈铁镂刻的字句，酝酿出不同风格的水乳交融、层次丰富的和谐空间。

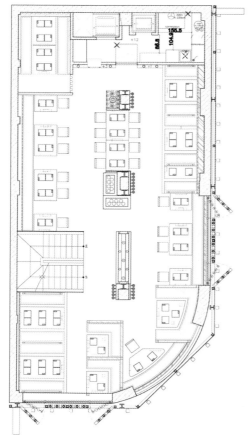
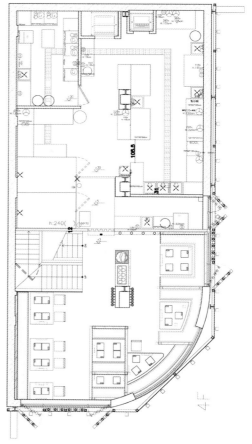

地点	面积	设计师	设计公司
/台湾	/400平方米	/周易	/周易设计工作室

Co-exist & Harmonious
共生 · 和谐

JOY Interior Design Studio makes good use of original building's congenital condition such as high interior ceiling room and floor to ceiling continuous windows and use an open field of vision and philosophical, and harmonious logical thinking to bring the Suzhou gardens like quiet indistinct of water, bamboo grove and narrow footway from external view into indoor box view, it also integrate modern classic black-white fashion contrast and Eastern culture fashion vocabulary and change and reconcile according to various proportions. Through the conflict and harmony between the two, it writes a whole new comment on the feeling of a pleasant dinner gathering.

Walk by the outdoor rough railroad tie, one side has a long line of square trail light with grid square window totem; elegant light and shadow tell the sincere welcome. Between the footway and the main building, a water view area is planned, in the borderless pool built with black stones, there are four huge black ceramic vats and candle lights and light beam springs which seem like floating on the pool are used to decorate the pol. In the center of the pool, an exposed form structure which stress minimalist aesthetic lines make appropriate connection and barrier between indoor and outdoor. Though the nature elements such as the wood, stone, light and water which were originated from the nature, the designer delivers the ultimate attainment between artificial technics and nature and refine the beauty of environmental harmony through quiet and comfortable environment.

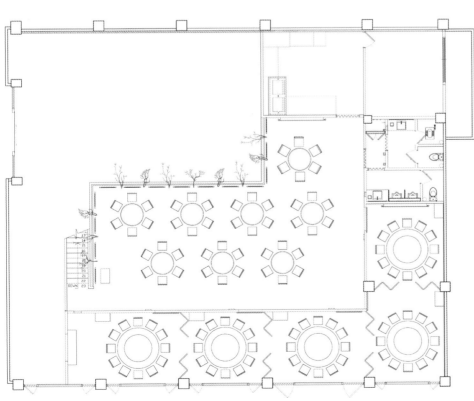

东方韵律 新中式餐饮空间设计

When entering the door, one could immediately see the extraordinary counter area. The advantage of 7 meter high ceiling in some area of the building is properly utilized, in the scene behind the counter white relief official script made with complicated process and almost perfect print typography skills transform the poem made by Li Bai with chic and lofty heroic feeling under the exquisite focus light into the three-dimensional visual art with strange momentum to decorate the words; the same chic red inscription with different size, and the conduct of the whole picture, in addition to implication with human virtue, is an unique and concrete practice of aesthetics.

The interior of the restaurant is using the fashion white-black contrast as the main tone, but the styling vocabulary is interspersed with a number of fragments of ancient China, for example, the lower edge of the dark staircase, a mirror like pond just like a replica of outdoor water views. A white dead branch in the pond has not only the "splash-ink painting" artistic conception of mountain and rivers, it also has the interesting scene of Japanese mountain and water, especially after the designer-Chou Yi who is always good at utilizing lights to decorate and make the goodness of life concentrated the view in front of our eyes. The first floor is mainly the open dining space, the beautiful scenery of bamboo forest outside the window is introduced into the interior though the French windows which greatly boost the appetite an emotional appeal.

The design of private box is also full of clever thinking, the white bamboo canopy and light shadow decorated on the edge of the wall make us associate with the laid-back feeling in a famous poem. Entrance door style is using the idea of elegant bottle; it has the sophisticated temperament of Chinese-style garden. The private boxes are named after famous Hakka community, it also points out the owner's idea of not forgetting ones origin.

环境内外本该是一个相互呼应的有机体，但由于考虑到安全、安静等因素，许多建筑物设置硬性阻隔，往往将室内、户外切割成两个界线分明的区块。不过在这个坐落于台湾的大型商场设计中，设计师充分利用原建筑物室内挑高与大面积落地窗的先天条件，以一种开阔的眼界、圆融的逻辑思考，将苏州园林般的流水、竹林、栈道等景观引入室内，并融入了现代经典的黑白时尚与东方文化语汇，依各种比例加以变化、调和，为令人心情愉悦的用餐意境写下了全新批注。

这栋仅两层楼高的独立建筑，有清水混凝土的质朴、陶缸水景的写意，还有玻璃帷幕的通透性。在鲜明的建筑语言里，同时注入了业主期许的自然、休闲概念，让人第一眼就留下了生动的印象。

走过户外粗犷的枕木栈道，单侧排布了有着格栅窗花图案的步道灯，以典雅的光影诉说着迎宾的热诚。栈道与主建筑物间有水景区，以黑色石材砌成的无边界水池里，点缀着四座巨大的黑色陶缸，以及看似漂浮的烛台灯与光束涌泉。

水池中央一座强调极简线条美的清水混凝土结构，让室内外有了适度的衔接与分隔，这些源于自然界的木、石、光、水等元素，传递出一种人造工艺与自然共生的景致，提炼出宁静与安逸的环境和谐之美，让所有到访者都能以最放松的心情入内用餐。

顾客一进即见气宇非凡的柜台区，设计师充分利用建筑物局部挑高逾七米的优势，在柜台后靠端景处以工序繁复的白底浮雕隶书和分毫不差的铅字排印技巧将李白的一首潇洒、豪情万丈的《将进酒》，在精妙的聚焦灯光下，转化为气势万千的立体视觉艺术。而点缀其中的大小不一却同样别致的红色落款的设计灵感来自乾隆皇帝喜欢在钟情的书画上捺印，隐喻专属的独特性。整个画面除了体现意蕴其中的人文涵养，更是一次独到美学的具体实践。此外，文字端景墙前精选黑色石材打造柜台主结构，立面以错落的实木块传递不经修饰的自然感，上方一排利落的铁架运用纤细的钢骨深入顶棚内强化支撑，再架上一长排烛灯，设计师因此将千年的文化与丰富的休闲景观尽情展现出来。

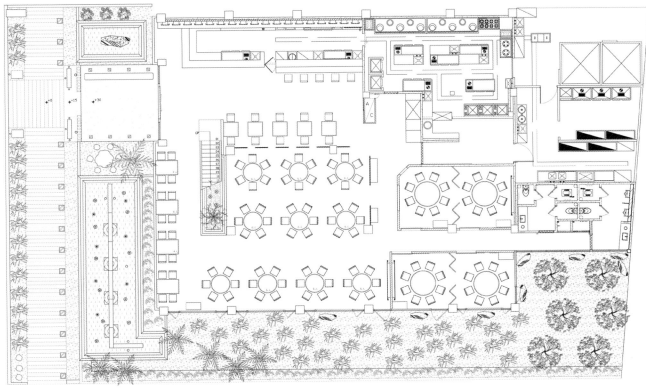

餐厅内部以时尚的黑白对比为主调，造型语汇上穿插了许多古老中国的片段。例如漆黑的楼梯下方，一方镜面旱池仿如是户外水景的复刻版。池中一棵全白枯枝，既有白山黑水的泼墨意境，也有日式枯山水般的耐人寻味。灯光的烘托将生活的无限美好浓缩于眼前的方寸之间。一楼主要为开放的用餐空间，其中一侧运用"有景借景，无景则避"的技巧，将落地窗外优美的竹林景致引入室内，大大提升了食欲和情绪的感染力。不容错过的还有点缀在窗畔的白墙与特定包厢内的大幅书法，这些全是当代知名书法家李峰的作品，其中一幅名为"如易"，很巧合地将设计师与业主名字中共有的"易"字带进来，营造出既有象征性又意义深远的艺术氛围。另一侧夹层上下使用大量中式窗花界定空间，全数喷白的线条格外立体，一字排开，颇具气势和震撼力。二楼顶棚处还有黑色枯枝迤逦而出，串联了空间，呼应了设计主题。

私密包间的设计同样极富巧思。在隔墙上方点缀的白色竹檐与灯光阴影，让人联想起"采菊东篱下"的悠闲，以雅致屏风为主的入口造型，有着中式园林的书香气质，门上的厢名则以知名的客家聚落命名，这也点明了业主的初衷。

地点	面积	设计师	设计公司	主要材料
/台湾	/1490平方米	/周易	/周易设计工作室	/铁件格栅、仿清水模瓷砖、橡木染黑、竹管、风化木、南非花梨木

KARUISAWA Tainan Store
轻井泽台南店

Located along the roadside," KARUISAWA" Tainan store has 30 meters wide store front, It is hard to imagine the store is the transformation of an old iron furniture store and is made into a landscape artwork. Rusty metal color contour lines are pulled on the top which makes the building naturally exhibit stability.

One large and one small water view with Zen prospect on the two sides of the arcade, each one has its unique features, the main water view on the left is like a long plate placing at high position, and the plate is embellished with stones, which showing the feeling of garnishing food presentation in Kaiseki style cooking. Gurgling flow of water on the surface of the plate is accomplished by beautiful lights, towering rocks seems to float on the water, the main feature for the water viewer on the right is plain and simple trunk.

The main dining space are concentrated in the flat first floor, a dry view of lights roughly shaped like two rectangles with the smaller one inside surrounding the center, there is a fence made with bamboo hanging in the air corresponding dry landscape with to two view rocks and Dohyo-like flat. The back section seats are nearby a huge glass windows, the area between the windows and the adjacent buildings is fully planted with bamboo trees, from the green bamboo forest in the back, the Dohyo in the middle to the water view and plants in the front, each view is connected with each other which has greatly enhance the interest and depth of "eating".

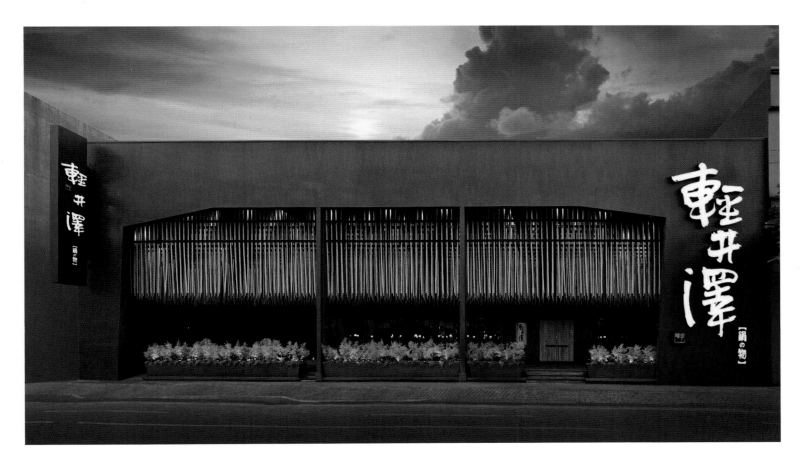

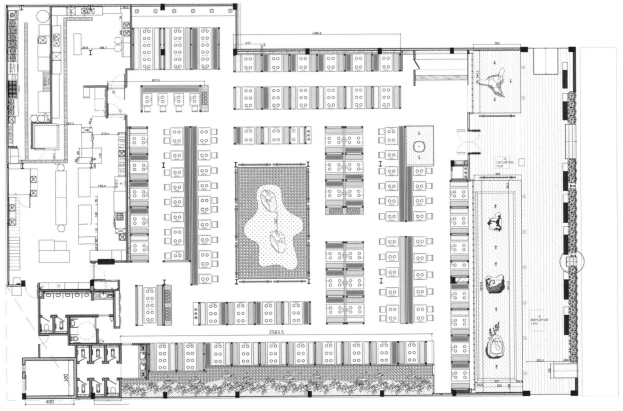

坐落于大道旁的轻井泽台南店宽为30米，很难想象这是由老旧铁皮家具卖场改造而成的。建筑顶部拉出水平线条的锈色金属轮廓，让建筑自然涌现出安定与稳重。

进入室内，两侧是一大一小、各拥奇趣的禅意水景。左边主水景宛如托高长盘，盘上点缀三块景石，颇有怀石料理摆盘的意境。盘面潺潺流动的水幕衬以唯美灯光，峥嵘奇石仿佛漂浮其上。右边水景则以朴拙瘤木为主角。

主要用餐空间都集中在一楼，大致呈"回"字形环抱。半空中由竹子排列而成的围篱，对应下方两座景石和枯山水景观。后段的卡座挨着大面积玻璃窗，窗外与相邻建筑间是生机盎然的竹林。从绿油油的后景竹林、中景的枯山水到前端的水景、植栽，环环相扣的景色大大提升了"食"的乐趣与深度。

地点	面积	设计师	设计公司	主要材料
/福建福州	/835平方米	/程济恒、陈振格	/济恒设计事务所	/壁纸、铜板、复合木地板

Yonghe Club No. 33
雍和会 33 号

This case in located inside an ancient building which used to be the landmark of the historic city of Fuzhou. The designer retained the building's vintage look but integrated some fashionable modern architectural elements into it, bringing a graceful dining experience to customers by taking advantage of its superior environment.

In the open courtyard, the designer did not paint the entire space, but kept the rough red bricks and mottled wooden structure, which show a LOFT style. In addition to retaining the original structure and materials, the design also used dining elements very creatively, putting colorful dishes on white walls and wine bottles on the crystal suspended ceiling. People would mistake them for pure modern-style decorations without closer examination.

The historic old items make the place look like an art museum, bringing people back to ancient times in an instant, making you feel strangely nostalgic. In order to reduce the differences between Chinese and Western elements, the designer paid close attention to replicating standard historic marks in the interior design of both the public area and the boxes. Visible wooden beams and pillars are not only the main supporting structure but also decorations. Though without exquisite carvings, they have created a plain and quaint style with primitive mortises and tenons.

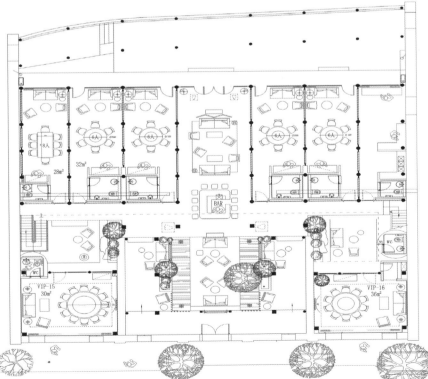

本案位于一座古宅内,这座古老建筑也是福州历史文化名城的标志性建筑。设计师保留了其复古的造型,融入了部分时尚的现代建筑元素,充分利用优越的环境优势,为食客带来高雅的就餐体验。

开阔的院落内,设计师并没有将空间内全部粉刷一新,而是保留了粗糙的红砖和斑驳的木质结构,流露出些许LOFT的味道。除了保留固有的建筑结构和材料,设计师还创造性地运用餐饮的元素,例如回廊里悬挂在白墙上的彩色碗盘、放置在水晶吊顶上的酒瓶,若不仔细看,还以为是现代风格的艺术装饰品。镌刻着历史感的老物件将空间装扮得犹如一个艺术展馆,将人们瞬间带回到那个遥远的年代,别有一番滋味。为了最大程度地融合中西方的文化,不论是公共区域还是包间的室内设计,设计师都特意复制那些具有代表性的历史痕迹。暴露在视线范围内的木质梁柱既是支撑建筑的主体结构也是空间的装饰。它们虽然没有精致的雕栏画柱,却用最原始的榫卯工艺营造出返璞归真的古朴氛围。

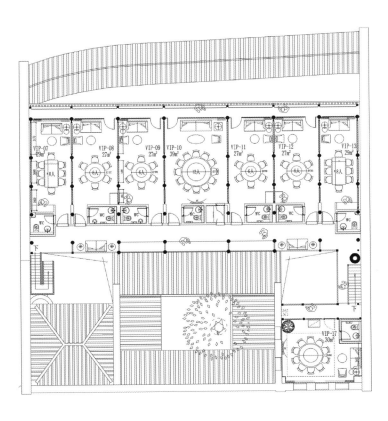

地点	面积	设计师	设计公司	主要材料
/台湾	/681 平方米	/周易	/周易设计工作室	/铁件格栅、梧桐木钢刷、杉木染黑、橡木洗黑、南非花梨木、茶镜

Elegant Leisure
寻幽竹篁里

In the process of going from the hustle and bustle of this world into the quiet course of time and space, the mind is gradually converted and eased, walk along the warm wood plank road under the foot, and then a mood of selflessness like the verses "Seated alone by shadowy bamboos, I strum my lyre and laugh aloud" appears in front of the visitors, a tray-like elongated elegant water feature is built, the pool is decorated with towering landscape stone, fountains, plants and exquisite trail lights are used to trim the edge of the walkway, showing the Kaiseki arrangement like elegant layout, an infinitely touching, sweet and chic gesture modern ink and wash painting like view is beautifully created.

The huge wooden structure aggregated through multi-level mortise, interpreting palace-like caisson imagery. The light-colored system which makes the body feel lighten is in stark contrast to the strong primary color in the environment, the wooden structure in the ceiling and the hand-made metal grid boundaries in the central seats, showing a both conflict and harmony ninety degree structure perspective, making people sincerely feel the vastness and forceful nature of space. Fodder barrels are hanged over alongside the road, rounded red persimmons are put inside the barrels, they are the richest flavor season fruits and they are used to proclaim that the store is opening its door for business, visitors and all the best wishes are sincerely self-evident.

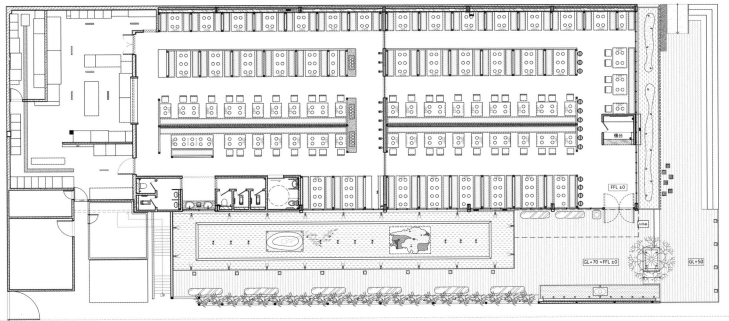

步入本案空间，顾客在从喧嚣尘世进入静谧时空的过程里，逐步获得适度转换、沉淀。沿着脚下温暖的实木栈道前行，眼前就是一番"独坐幽篁里，弹琴复长啸"的忘我意境。设计师打造出宛如长形托盘般的飘逸水景。池内缀以峥嵘景石、灯光喷泉与植栽，并以精致步道灯镶边，颇有怀石料理摆盘的风雅布局，放眼望去一幅风情无限、姿态潇洒的现代水墨嫣然生成。

通过多层次连接聚合的巨大木结构，诠释犹如殿堂般的藻井意象，使轻盈的浅色系与环境的主色调的浓重形成强烈对比，而顶棚的木结构与中央卡座分界的手工制作的金属格栅既冲突又和谐，令人感受到空间的浩瀚雄浑。过道旁悬挂着做旧的木桶并一字排开，内置成堆珠圆玉润的红柿，堪称最富和丰收的季节，昭告店家开市大吉、顾客事事如意的诚挚祝愿尽在不言中。

地点	面积	设计师	设计公司
/福建福州	/750平方米	/林鸿	/维野商业空间设计

Lushe Dining Club
麓舍餐饮会所

The project is located at the foot of mountains full of trees, where it is very beautiful with pleasant climate, a quiet and comfortable hidden world. The natural and pure space atmosphere represents the unique characteristic, so named "Lu homes". As for the design, traditional Chinese elements are strictly selected and applied to every space of the club just right.

There are totally five boxes, one book room and three tea rooms in the restaurant. The designer attaches more importance to the function without more gorgeous decoration on its configuration, full of Chinese elements, real, pure, elegant and exquisite on dominant hue.

People pass by classical passageway paved by grey bricks and bluestones to a hall. The hall is not splendid, but represents a sense of simple and free comfort. The application of classical woods, hessian, stones, bricks and white wall makes the space more modest. Traditional Chinese ink paintings are combined with Fuzhou bodiless lacquer ware of Chinese traditional artifact "Top Three", root carving as well as lacquer painting, to add more cultural atmosphere to the space, which touches customers' hearts and inherits the oriental culture.

The suspended ceiling in the box is designed elaborately. Chinese traditional wooden ceiling and dining tables are under the chandelier's shadow, to make people seem to be in a ancient house. The suspended ceiling in the small box is made by knitting boards, to be more modest and pleasant in the space, so people inside also enjoy the elegant environment. Chinese style gradually fuses with modern aesthetic to be closer to real life, to keep a gentle and kind traditional feeling.

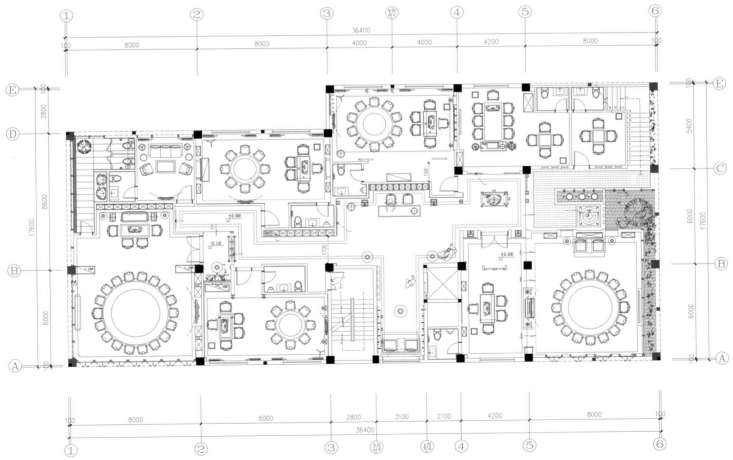

东方韵律 新中式餐饮空间设计

本案位于山林麓间，环境优美、气候宜人，是一处安静舒适的隐世之所。其自然淳朴的空间情境让餐饮氛围拥有了别样的气质，故取名"麓舍"。设计将传统的中式元素经过严格的筛选，恰到好处地运用于会所的各个空间。

餐厅共设有5间包间、1间书画室和3间茶室。设计师在设计中更加强调功能，装饰造型上没有过多华丽的元素，中式笔墨挥洒其中，彰显出真实、纯朴，整体色调古朴雅致。

穿过青石、灰砖布置的古色走道，便进入大厅。大厅门面并不气派华丽，却透露出简单、随意的舒适之感。仿古木、粗麻布、原石、青砖、白墙的材质运用让空间更加朴实。空间结合中国传统水墨画及福州脱胎漆器、根雕、漆画这些中国传统艺术品，给空间带来了更多的人文气息，既感染了宾客，又传承了东方文化。

包间的顶棚经过了设计师精心的设计，中式传统木架吊顶与餐桌在吊灯的光影渲染下，让人恍若置身于一栋古宅中。小型包间则是使用编织板来塑造顶棚，让空间氛围更加质朴怡人，营造出一个优雅的环境。该设计融合了中式风格与现代审美，更加贴近真实的生活，并保持着那份温厚的传统情怀。

地点	面积	设计师	主要材料
/福建福州	/900 平方米	/张清华	/波尔多灰泥、青石、木雕板、杉木板

Hakka Impression Traceability Club

印象客家溯源会所

The case is to show the Hakka cultural connotation of the original eco-Hakka-themed dining club, a high-end clubs which business position is based on Hakka original ecological food, and supplemented by Hakka native products, handicrafts, paintings. As the participation in club positioning before design, so at the beginning of the spatial layout, the whole design has keep to Hakka culture, by means of space architectural language Hakka Earth building, stress on the relationship between the vertical and horizontal lines, intersections, and also consider the natural circulation between space and air.

The material selection of space decoration and construction is also based on the original ecological environment protection materials, such as bluestone, brick, plaster, gravel rice, etc., and keep some of the original architectural wall granite slope. Not too much decoration, pure space, aimed at showing the Hakka spirit.

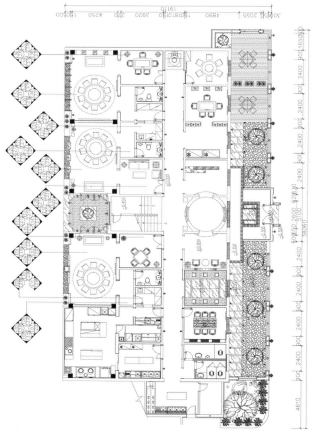
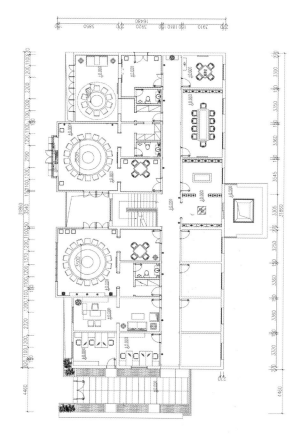

本案是以展现客家文化内涵为主题的原生态餐饮会所。经营产品以客家原生态美食为主，同时以客家土特产、工艺品、字画为辅。因为设计之前就确定了会所的定位，所以在一开始的空间布局中就遵循结合客家文化，借鉴客家土楼的空间建筑语言，在设计中讲究纵线与横线、交叉点的相互关系，同时还考虑空间的自然流通。

空间装修选材上也是以环保原生态的材料为主，如青石、青砖、灰泥、砂砾等，并保留一些建筑原始的护墙麻石。设计没有过多的装饰，营造了一个纯粹的空间，意在呈现出客家人的精神面貌。

地点	面积	设计师	设计公司	主要材料
/福建福州	/600平方米	/张开旺、林文施	/大于空间张开旺设计组	/钢板、旧木板

Lian Yugang Seafood Restaurant

连渔港海鲜餐吧

"Lasting by leading a quiet life" is the first impression given by this case. It was just like a hermit in the bustling city, who enjoyed the comfortable of quiet at the corner. The fresh green created a quiet and mild environment for dinner like a spring breeze, and when it echoed with the log surroundings, the sense of elegant of the poetic Jiangnan was brought into the space. The inspiration of space came from the Jiangnan windows which were cut off and endowed with rich impressions. As a result, the special verve of new eastern aesthetics was conveyed.

The usage of color for this case was very flexible: a fresh and naturally atmosphere was created with a keynote of light green and wood color; the colorful glass and art wall in the space not only energized the indoor, but also increased the beauty and made the performance effect more diverse. A little of lively elements in the quiet space would turn out more surprise. In addition, the lighting and wall hangings of this case were another highlight. The photo wall which had much stories, the lifelike decorations of deer head, shapes of iron arts and lightings, all of this expressed the good taste and style completely.

宁静致远是本案给人的第一感觉。客人犹如隐匿于熙攘都市之中的隐者，独享着偏安一隅的惬意。清新的绿色仿若吹入室内的一缕春风，营造出宁静、平和的就餐氛围，并与周边的原木材质相互呼应，将诗意江南的清雅、安然带入空间中。空间设计灵感来自于江南寒窗的隔断，它被灯光赋予丰富的表情，传达出新东方美学的独特神韵。

本案的用色极其灵活，以淡绿色和木色为基调，烘托出清新自然的整体气氛。

五彩斑斓的玻璃和艺术墙砖恰到好处地出现于空间中，不仅为室内注入活力，还增加了美感，使空间更加多元化。当宁静的空间中出现一些"热闹"的小元素时，就会给人带来更多的惊喜。

此外，本案所选择的灯饰和墙面挂饰，也是一大亮点。故事感十足的照片墙、栩栩如生的鹿头装饰、形态各异的铁艺灯具等，将空间的不俗品位与格调体现得淋漓尽致。

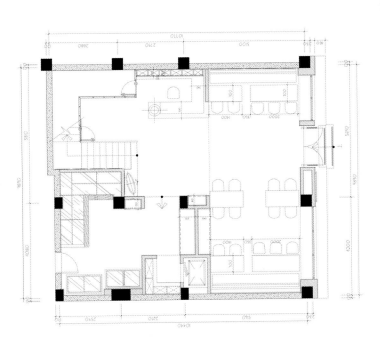

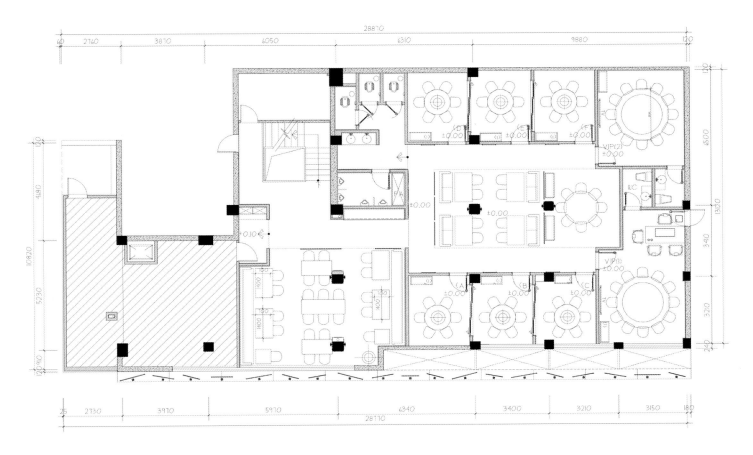

东方韵律 新中式餐饮空间设计

地点	面积	设计师	参与设计	设计公司	主要材料
/江西	/23 000平方米	/赵牧桓	/王颖建、赵玉玲、胡昕岳	/赵牧桓室内设计研究室	/实木雕刻板、漆、花岗石、工艺竹编、刺绣壁纸

Jiangxi Hengmao Resort Hotel's Chinese Restaurant

江西恒茂度假酒店中餐厅

The design of this case has drawn inspiration from Tao Yuanming's famous prose Peach Blossom Spring. As you enter the lobby, you'll see stone carvings of the lines from the essay on the wall and feel an ancient Zen state in the capacious room. The space is separated but not completely divided by hollowed-out screens, which blend perfectly with the traditional furniture and decorations. The pillars and decorative lights on the ceiling use two different materials—wood and metal, creating an tension in the space. Sharp metal and exquisite wooden lines emphasize the space structure while delicate pendant lamps show a unique charisma from inside out.

The ceiling design has restored the original structure of the roof and created an open space, allowing customers to breathe freely in the bright sun and clean air, and naturally integrate into the surroundings. The furniture and lighting inside are concise and plain with calm colors, bringing an elegant dining atmosphere. Chinese writing brush element was used in the ceiling lamps, demonstrating an extremely post-modern temperament. The Chinese-style carved window grilles and bookshelves in the large boxes have created a casual and leisurely environment.

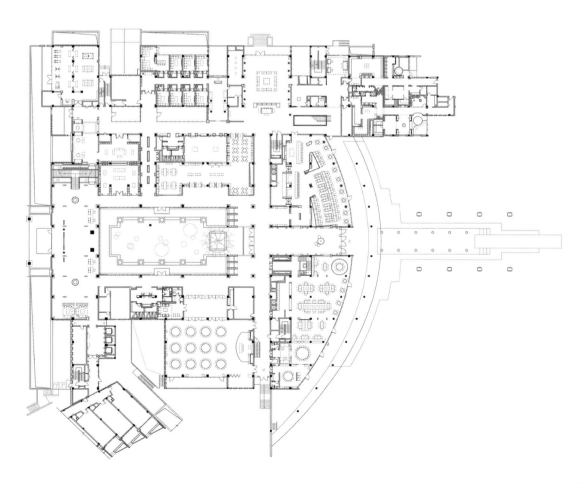

东方韵律 新中式餐饮空间设计

本案的设计灵感源于陶渊明的《桃花源记》，走入大堂，室内空间宽阔，墙面灰色石材上雕刻了《桃花源记》中的诗句，禅境古意扑面而来。镂空屏风的设计让空间隔而不断，与具有传统韵味的家具和配饰相互交融。柱子和顶棚上的装饰灯具用金属和木材两种截然不同的材料和质地制作，创造了空间的内在张力，并通过锐利的金属、细腻的木质线条强调空间结构。精致笔挺的吊灯令空间由内向外展现出独特魅力。

顶棚设计还原建筑屋顶原有结构，保持空间开阔，与周边环境自然融合，让顾客在明媚的阳光下自由呼吸纯净的空气。室内的家具、灯具，外观简洁质朴，其宁静的色调营造出优雅的用餐气氛。其间的顶棚吊灯，巧妙引入毛笔元素，具有极强的后现代艺术气质。大包间的墙面中式雕刻窗花的组合和书架造型，构建出一个轻松、随意的环境。

东方韵律 新中式餐饮空间设计

地点	面积	设计师	设计公司	主要材料
/北京	/1400平方米	/王砚晨、李向宁	/经典国际设计机构（亚洲）有限公司	/丝网印刷玻璃、硅藻泥涂料、黑色抛光地砖、黑钛不锈钢板

Nanxincang Branch of Dadong Chain Restaurant

大董餐厅南新仓店

The Dadong Chain Restaurant Nanxincang Branch lies in ancient royal granary district for Ming and Qing dynasties. As a project of expansion, this program includes a multi-functioned hall and four (4) top-grade separated dining rooms.

Bamboo has been concluded subject in this new district. Method of cavalier perspective broadly used in traditional Chinese landscape paintings, has been adopted in the whole program, various bamboos, including close, medium and distant shot, have been located in the same space, which forms magical layers and scene depth. Thanks to the cutting-edge technologies, advantages of Chinese wash painting have been demonstrated to its maximum, water and ink, black and white, all merging in a space with highly surrealism. The bamboo leaves, in the floor and walls, change color under LED control system, build different natural scenes and seasons. Most materials used are local and natural materials, recyclable, with knowledge and understanding of designers to traditional Chinese culture. We hope to convey graceful interests and charm with wash painting, and to build a classical space filled with pure Chinese culture and wisdom.

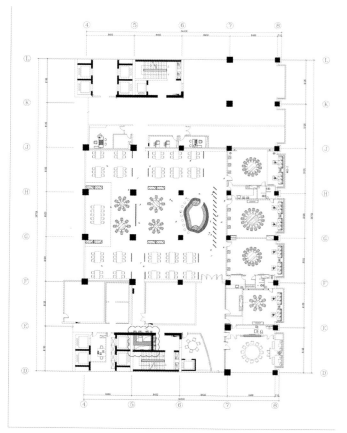
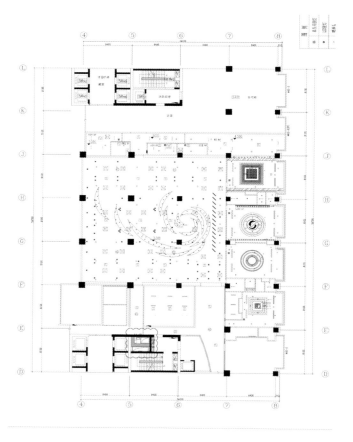

大董餐厅南新仓店位于北京明、清两代的皇家古粮仓群处。本项目作为新扩建的部分，包含一个多功能厅及四个高档独立包间。

竹是这个新区域的设计主题。中国山水画中的散点透视手法被运用于整个空间，竹的不同的近景、中景、远景同时融于一个空间之中，形成了丰富的层次和景深。最新技术的运用，将中国水墨艺术的淋漓和洒脱发挥到了极致。水与墨、黑与白在超现实主义的空间中互相渗透。墙面与地面发光的竹叶，通过LED照明控制系统来变幻色彩，从而营造不同的自然场景，让人感受四时四季的变化。使用的主要材料都是当地的和可再生的天然材料，并融入了设计师对中国传统文化的认知及思考。设计师希望用水墨来表达隽雅的趣味和韵味，打造纯正的中国式的、充满东方文化与智慧的经典空间。

地点	面积	设计师	设计公司	主要材料
/上海	/1058平方米	/甘泰来	/齐物设计事业有限公司	/铁件、软木地板、盘多磨、白洞石

DOZO Izakaya Dining Bar
DOZO 创作料理

The project is located in the center of the city where shopping malls stand and there are many of first-class brands as well as foreign enterprise office buildings nearby. The entrance of the restaurant is on the first floor, facing the building square, which specifically establishes the detour path based on the concept from "contraction" to "enlargement", due to crowded flow of people. Upon arrival to the second floor, people have to bypass the screen to enter the inner dining area, where a quiet and mysterious space suddenly turn to a broad vision and open seats on the second floor, resulting in strong contrast.

The space is divided into open area and boxes area. The counter area is planned in the front of the open area, offering a place for companionship and drink. Inside the space of six-meters high, stairs are made to build rich space levels. The maze garden is regarded as the theme of boxes area, where designers take the advantage of the boxes to make the walk-way like a path, and use "edge" design combined with external parts, where the grid door inside the boxes and elastic partition way from the shutter expand the activity area for guests who can interact with adjacent areas casually, and feel a cheerful atmosphere of friendship from each other.

本案位于商场林立的市中心，附近聚集了许多一线品牌及外商办公大楼。餐厅入口位于一楼，面朝大楼广场。由于人流密集，特意以由"缩"至"放"的概念建立迂回路径。抵达二楼时，需绕行屏风才得进入内部餐厅区，由安静神秘的空间前奏，转为二楼的开阔视野及开放坐席，如此设计产生强烈的反差对比。空间分为开放区与包间区。开放区前方规划了吧台区，提供了朋友小酌的场地。

在六米高的空间里，利用阶梯打造出丰富的空间层次。包间区以迷宫花园为主题，利用包间塑造出小路般的步道，并利用开放的设计手法与外部连接。包间内部的格栅拉门、卷帘隔断的区隔方式，扩大了宾客活动区域，可让宾客随兴地与邻区互动，感染彼此欢愉的氛围。

地点	面积	设计师	设计公司
/辽宁葫芦岛	/2101 平方米	/赵睿	/大连纬图建筑设计装饰工程有限公司

Huludao Food House Private Dining

葫芦岛食屋私人餐厅

"Meal house" is defined as a private club for owners and their relatives to gather for a party, not concerned with commercial purpose. Therefore, there is no need to deliberately cater to public taste, and there is a relatively free space for the designer to take a "process" creation and to thoroughly release their mood.

The predecessor of the project is a restaurant open to public, whose architectural style is the typically retro architectural trend in the 1970s and 1980s. Geographic position is great with broad vision and directly facing the sea scenery from windows. The designer starts the project from external walking way to internal walking way. Big entrance steps originally facing to the roadside are removed and placed on the west side of the project, from which people walk up, and stone lions symmetrically placed on both sides add a lot of fun and a sense of ceremony to the entrance. During the walking way of a sun terrace reconstruction, the designer tries his best to avoid existing plant landscape, to make construction plane represent a sense of belonging in the site, to be more "natural". The project appearance is reconstructed with dominant hue of gray and white, and displays itself with trapezium shape, which is integrated to the surrounding scenery, co-exist harmoniously.

The designer applies something like "straw" to decorate the whole "meal house" space, and then "straw" items appear at the entrance of ante-hall, as well as the walls between spaces and the ceiling full of irregular wooden texture. The Designer hopes corners of the ceiling and the same style spread in three-dimension, and improvises on his design of the site, to extend the space to a certain extent, more of artistic creation of unknown exploration.

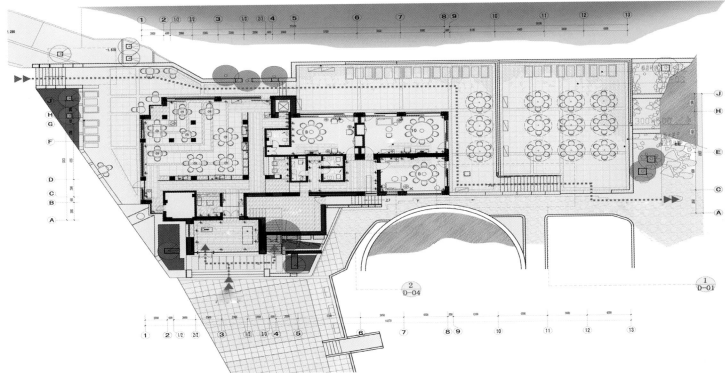

东方韵律 新中式餐饮空间设计

"食屋"定位为私人会所，供业主和亲友在此聚会使用，并不考虑对外商业用途。因此无需刻意去让室内装饰迎合众人，这也为设计师提供了一个相对自由的空间来进行"过程式"的创作和自我情绪彻底的释放。

本案前身作为餐厅是对外营业的，其建筑形式为典型的七八十年代复古建筑风格。地理位置优越，视野开阔，并且可直面无边海景。设计师首先从建筑内外动线展开设计。原本直对着大马路的入口大台阶被移除，并设置在建筑的西侧。拾级而上，两侧对称分布的石狮为入口增添了不少趣味和仪式感。在户外阳台的增建过程中，设计师尽量避开场地现有的植物景观，使建筑平面轮廓的形态多了一份"自然而然"的场地归属感。经重新改造后，以灰白色系为主的建筑外观，使得建筑与周边景观融为一体，相得益彰。

设计师把"稻草"置换成一种语言融入到"食屋"的空间设计中，进而出现了入口前厅的"稻草"装置，以及在每个空间节点的墙面和顶棚上延续的不规则木条肌理。设计师希望基于这样的装置节点设计，再加上设计师的现场即兴创作，让空间创作的边界在一定程度上延伸，使空间多一些艺术创作的未知性和探索性。

地点	面积	设计师	方案审定	设计公司	主要材料
/福建	/500平方米	/庄锦星	/叶斌	/福建国广一叶建筑装饰设计工程有限公司	/毛石、青砖、芝麻灰火烧板、仿古砖

Yiran Ju
怡然居

Red tilt and green bricks shines, bamboo shadow leans from the window, all reminds you the pictures of Jiangnan. Inspired by delicacy and elegance of new Chinese style, designers in this case create this quiet dining space and endow new definition with catering culture.

Designers decorate the space with typical oriental elements, suffusing the space with ancient charm, traveling the millennium. Green brick schema the whole space, rendering out years of traces. Window grid as partition, chic and fine, delivers constant exquisite feelings, making the room spacious; Storage real cabinet, chairs and floor made from logs, natural and fresh, form a poetic picture with the dotted bamboo; Preface to the Collection of Poetry at Lan Pavilion matches the stone lions, revealing talent and elegance of Jiangnan. In addition, designers select furnishings by heart: lightly rolled curtain, low-key porcelain pervades, artistic decorative painting, simple and unique decorative lighting quietly tell their own overwhelming story.

青砖黛瓦，竹影横斜，小轩窗，梦回画意江南。本案的设计师用新中式风格的清浅雅致，打造出了一个宁静的就餐空间，赋予了餐饮文化新的定义。

设计师采用了大量具有代表性的东方元素来装饰空间，让室内弥漫浓郁的古典韵味。置身其中，仿若穿越千年。古朴的青砖架构着整个空间，呈现出岁月的痕迹。作为隔断的窗格，别致精美，将淡淡的思古幽情传递出来，并有着隔而不断的效果，增加了空间开阔性。原木质地的储物柜、桌椅与地板透露出自然清新的气息，点缀其间的翠竹构成了一个个富有诗情的画面。肆意挥洒于墙面的《兰亭序》与周边的石狮交相辉映，将江南的才情与风雅尽情展现。此外，陈设品的选择也体现出设计师的用心。轻卷的竹帘、低调的瓷器随处可见，意境独特的装饰画、简约个性的灯饰静静地诉说着属于自己的故事，让人不禁为之动容。

东方韵律 新中式餐饮空间设计

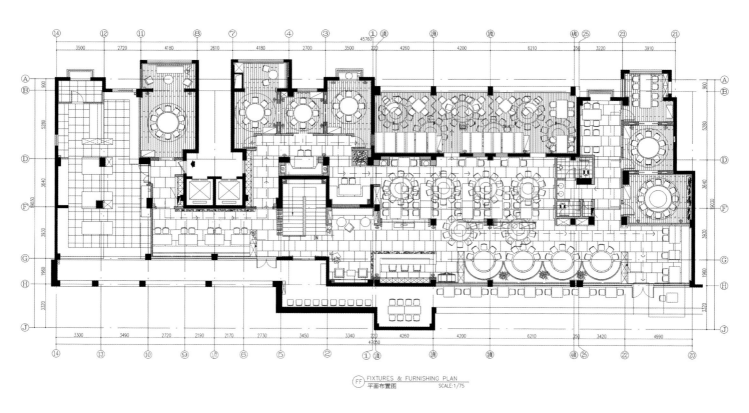

FIXTURES & FURNISHING PLAN
平面布置图　SCALE:1/75

地点	面积	设计师	设计公司	主要材料
/湖北	/1900平方米	/王砚晨、李向宁、郭文涛	/经典国际设计机构（亚洲）有限公司	/大理石、生锈钢板、橡胶木、金属网

Wong's Hot Pot Restaurant
王家渡火锅餐厅

Water and Skylight

Wong's Hot Pot Restaurant is located in the central region of Yiaihu Park of Huanggang, Hubei. The Yiaihu Park is the most beautiful natural landscape lake park in Huanggang. In the park, there are hanging willows sweeping the lake bank. The sparkling water in the lake looks like elegant silk with sense of poetry in its folds. The grand lake is clean and crystal clear, and the lake waves and willows reflect each other, bringing endless tenderness to people.

Walking along the lake path to the restaurant, you can enjoy the natural beauty of the lake just as Su Shi's (a famous ancient writer) wonderful description in his poetry of the West Lake scenery. In sunny day, you can enjoy the water, and in rainy day, you can also enjoy the mountains. Surrounded by landscapes of water and mountains, the environment of the restaurant brought design inspiration. Through reasonable retention and use of the surrounding plant, redefining the relationship between architecture and nature, balance of design and nature is achieved.

The interior space design concept derived from the core brand idea of Wong's Hot Pot Restaurant — the re-interpretation of ferry crossing culture. Therefore, the human and the natural elements related to ferry crossing culture enriched the space design language. Water lines, pebble, fish, waterfowl, calamus, wupeng boat, plank road and other visual images were treated with abstraction refinement, being expressed with different materials. In the space, traditional materials such as metal, glass, stone, wood became the new carrier, with innovative technique weaving together a carefree pure beauty of natural scenery.

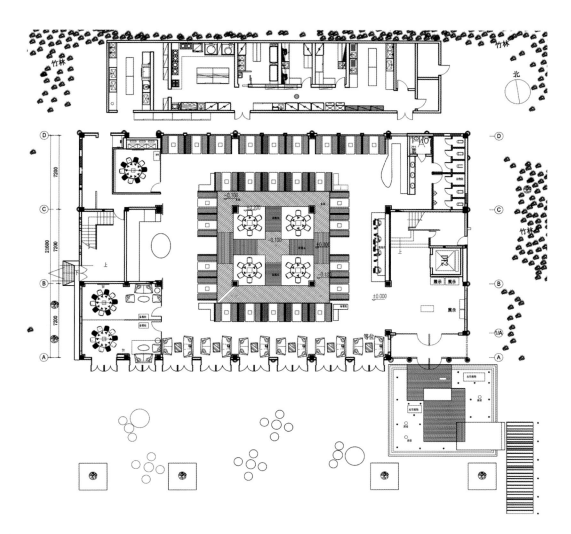
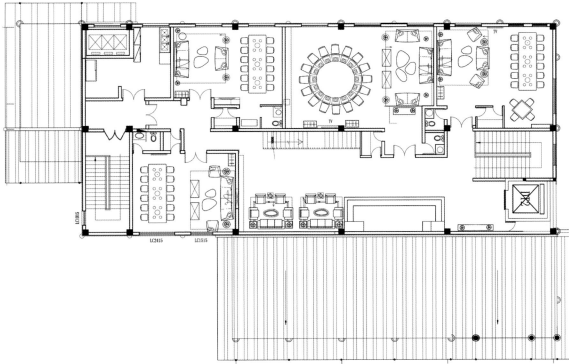

The sightseeing balcony in the top floor is a best place to enjoy the lake views. The wooden platform, jade-like white fence, and green tiles form a private space, with the broad vision offering infinite possibility to enjoy the lake views. Whenever in the morning or at dusk, sitting at the terrace, bathed with the wind, beside the fence, you can enjoy the feeling as described by Su Shi: This remind me of the gleaming mountains of a beautiful poem. Broad water like a mirror, clean and clear, with green peaks reflected.

水色天光

本案位于湖北省黄冈市的遗爱湖公园腹地。遗爱湖是黄冈市内最美的城市自然湖景公园。湖畔种植的大多是垂柳，湖面波光粼粼，如丝绸般飘逸，褶皱处也满含诗情。广阔的湖面明净而通透，湖波如境，杨柳夹岸，照映倩影，充满无限柔情。

沿着湖边小径走向餐厅，湖边的自然美景恰如苏轼描写过的醉人西湖景色："水光潋滟晴方好，山色空蒙雨亦奇。"隐逸于山水之中，这是餐厅所处环境对设计的灵感启发。通过合理保留与利用周边植栽，重新定义建筑和自然的关系，达到设计与自然的平衡。

室内空间的设计概念源自王家渡火锅的品牌核心理念，即渡口文化的重新演绎。于是有关渡口文化中的人文和自然元素演变为空间中的设计语汇。水纹、卵石、游鱼、水鸟、菖蒲、篷船、栈道等视觉意象通过抽象化提炼，以不同的材质来体现。在空间中，金属、玻璃、石材、木材等传统材料成为新的载体，以创新的手法共同编织一幅悠然、纯美的自然美图。

顶层的观景露台更是欣赏湖景的绝佳之地，木质地台、白玉围栏、青黛瓦面围合成一处私密的顶层空间，提供了欣赏湖景的开阔视野，无论是清晨还是日暮，坐在露台上，沐浴清风，凭栏远眺，无比惬意，正如东坡词所说："认得醉翁语，山色有无中。一千顷，都镜净，倒碧峰。"

地点	面积	设计师	设计公司	主要材料
/北京	/480 平方米	/吴为	/北京屋里门外 (INX) 设计有限公司	/青石板、水曲柳、灰麻花岗石、木地板、麻刀灰、草编壁纸

Yiyun Xiang Restaurant
溢云香餐厅

This case is a restaurant space operating Yunnan cuisine, in front of such a national distinctive restaurant, designer integrated Dai, Bai, Yi, Tibetan and other national art as a premise, established rough, natural space design tone, allowed with a rich Yunnan flavor.

All furniture in the restaurant are designed specifically for the space by designer, original flavor wooden tables and chairs matches with a striped pattern of ethnic cushion, rough texture beams with transparent partition, abundant of green plants with obvious signs of manual pottery …… All these seem to remind diners they are living in the south of cloud district.

To emphasize the natural elements of space, plant wallpaper are widely used on walls, rough and full of sense of texture; Moreover, walls also presented with more "transparent" — not completely closed solid walls, each wall are artificial "cut" to put plants; and gable roof of the individual area also appears to remind diners this is a Dai "bamboo house," ventilated everywhere.

In such a national style space, designers deliberately avoided a lot of obscure national elements instead of a popular territory symbols, making those who have or never been to Yunnan able to savor to the enjoyment of the body and mind brought by their space since step into.

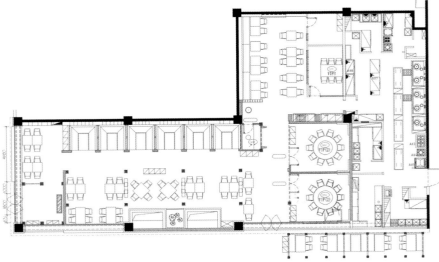

本案是一家以经营云南菜为主的餐饮空间，面对这样一个拥有民族特色的餐厅，设计师在结合傣、白、彝、藏等民族艺术的前提下，确立了粗犷、自然的设计基调，使空间带着浓郁的云南风情。

餐厅中的所有家具皆为设计师专门为这个空间设计的，质朴的木质桌椅搭配条纹图案的民族坐垫，具有粗糙质感的水曲柳房梁搭配通透的隔断，盎然的绿植搭配手工痕迹明显的陶罐……这一切似乎都在将食客的身心带到彩云之南。

为了强调空间的自然元素，墙面多用植物壁纸，粗糙且富有肌理感。另外，墙面也多以"透"来呈现——没有完全封闭的实墙，每一面墙都人工"开凿"，用以摆放绿植。而散客区的"人"字形屋顶也似乎在提示食客这是一间处处通风的傣族"竹楼"。

在这样一个民族范儿的空间中，设计师刻意回避了很多生涩的民族元素，取而代之的是大众化的地域符号，让去过或没去过云南的人，从进入空间开始，便能细细品味空间为其带来的身与心的享受。

地点	面积	设计师	设计公司	主要材料
/北京	/1050平方米	/吴为	/北京屋里门外(INX)设计有限公司	/老榆木、青石板、大理石、麻刀灰、红砖、砖雕

Seasons Blessing of Duck Restaurant, Mongkok

四季民福烤鸭店旺角店

The project is always defined as tradtional dishes of old Beijing, mainly for toasted ducks. It is very popular with good remark upon selected materials and creative flavors. This is what the designer builds the space for. It is to taste traditional innovation. Upon the previous Chinese style restaurant design, some classical patterns of traditional architecture are applied on the external space. In the project, outside the elevator door is decorated by traditional richly carved arch over a gateway, with classical elments into it as well as traditional classic wrapped by modern, to convey a contemporary Chinese aesthetic in this way.

Open-plan toasted duck room is placed in the ante-hall, as the process of toasted duck procuced by cooks is a strong visual art, which represents a slow operation of toasted duck making. The design theme of "a slow life" is an accurate defination to the project set up by the designer, and a breakthrough point at the same time. The designer adopts a very natural way to express the original ecology of the space by materials with natural texture as well as the most modest form, to show a kind of life.

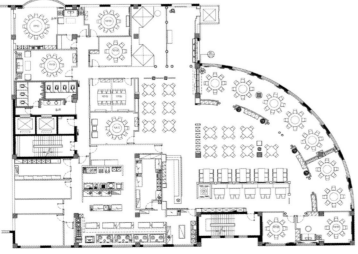

本案是以烤鸭为主的老北京传统菜品餐厅，因用料考究、口味创新而颇受好评。这正如设计师对本案空间设计的定位，在寻常之中品味传统的创新。以往中式风格餐厅设计中，经常会将一些传统建筑中的经典形式应用于空间的外立面上。在本案的设计中，电梯门的外侧使用了传统的雕花门楼作为装饰，将经典元素置于一个简约的盒子中，用现代包裹了传统的经典，以这种方式表达了一种当下的新中式审美。

开放式烤鸭房设置在了前厅，烤鸭师傅的操作过程具有很强的观赏性，呈现出烤鸭的慢制过程。本案以"慢生活"为设计主题，这同时也是一个切入点，用自然的方式来表达空间的原生态，通过自然质感的材料，以最朴实的形式展现出一种生活的状态。

地点	面积	设计师	设计公司	主要材料
/广东汕头	/600平方米	/叶晖、陈坚	/汕头市今古凤凰空间策划有限公司	/银白龙石板、英国蓝石板、黑色拉丝不锈钢、水曲柳实木、工艺纱

The Taste of Chao Shan
潮汕味道

This case is a Chinese restaurant extremely characterized by its special Chao-Shan taste. The aim of the design is particularly clear. The restaurant consists of three parts including private seats, extra seats and box area.

The Chinese landscape painting on the wall of the reception hall extremely conforms to psychological characteristics of the modern people who want to find a quiet and comfortable environment. The blue and white porcelain and lamp on the showcase, the display of the large cage in the stair, the decoration of the delicate lotus and goldfish medallion on the wall, and the special lighting on the ceiling, all integrate as a unique scene, making the classic lasting appeal of Oriental culture among them. Simple and crafted wood flowers and screen is indistinct, making the whole dining area and the box area share the same space, and that will give you a fresh aesthetic experience. The brief ceiling and the fresh wooden floor in the dining area integrate modern style with something classical, making dining-room full of Oriental culture but also giving a person a kind of comfortable and relaxed feeling. The decoration methods of traditional Chinese classical elements combining with the modern material texture bring the popular and classic element together, giving a guest the full feeling about the delicious dinner.

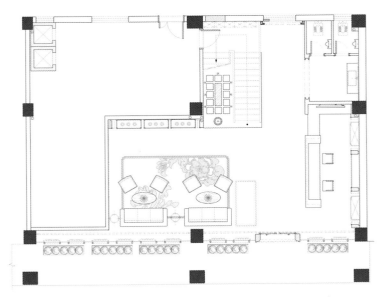
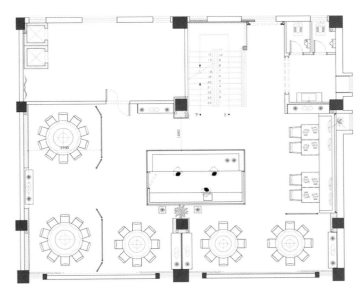

本案是一家极具潮汕味道的中式餐厅，室内设计层次分明、错落有致。餐厅包括雅座、散座大厅、包间三个部分。

入口接待大厅墙上的水墨山水画，极其符合现代人想找一个安静、舒适环境的心理特征。展柜上的青花瓷瓶和台灯、楼梯间的大型鸟笼摆设及墙上精致的荷花和金鱼挂饰、顶棚上鸟笼般的灯饰等相映成趣，浑然一体，使具有东方文化的古典韵味弥漫其中。线条简约、精雕细刻的实木通花屏风使其后面的空间若隐若现，使整个用餐区与包间具有空间共享的效果，给人全新的审美体验。用餐区简约的顶棚及简洁清新的木地板地面，让现代感融入了古典之中，使餐厅在充满东方文化底蕴的同时又给人一种舒适、放松的感觉。传统的中国古典元素和装饰手法与现代材料质感的结合，让流行与经典同列一室，构成了新的概念、新的视觉，使客人在就餐时充分享受美味。

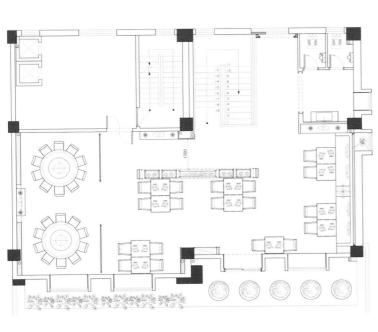

地点	面积	设计师	设计公司	主要材料
/河北唐山	/1500平方米	/吴晓温	/大石代设计咨询有限公司	/灰砖、水泥砖、橡木、涂料、水泥、金属网

Donglaishun TianYuan Restaurant

东来顺天元店

The design concept of the project is defined as "market living hall", whose characteristic is being rich, lively and full of scene, experience, civilians and fusion.

This project is interpreted by post-modern technique with traditional local flavor. "Feeling meal", namely a heart taste, "market culture" is the public consumption, which has a strong sense of intimacy, a sense of excitement and a sense of experience. The social attributes of urban life has been far away from people's natural attribute. Leading nature to life and the concept of market and street runs through the whole dining process. Doorway is the decorated archway of market, in the front of which displays wooden boxes where an episode of market scene is painted on each of them and metal cages with various sizes, building a dazzling market scene in the space. The passageway and the corridor are full of wheel print from wheelers, the real wheeler is at the display wall place, and plant walls with large area bring vigor and vitality to the whole space. Different street sceneries reappear in all rooms on the two floors.

Art and life are not supposed to be opposed, so it should be art from galleries and museums to the public life, art close to civilians. Art is supposed to be expressed by a friendly, such as reconstruction of old buildings, old furniture and materials representing a new vitality by the second design, full of passage of time, stories, memories, environmental protection concept and eternal spirit.

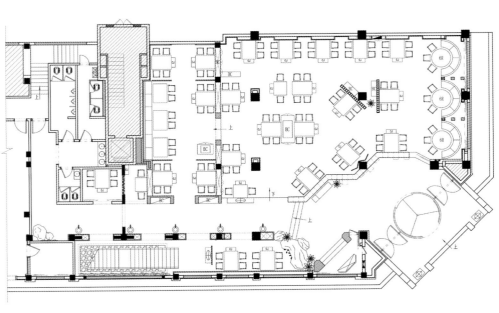

本案的设计主题定位为"集市生活馆",其特点是场景化、丰富化、热闹化、体验化、平民化、融合化。

本案用后现代的手法演绎传统的乡土气息。"感餐"即心里的味道,"集市文化"即大众消费,空间具有很强的亲切感、热闹感、体验感。远离都市生活的社会属性,将自然引进生活,把集市、街的概念贯穿在整个就餐空间中。门头是集市的牌坊,罗列了大大小小的木箱和金属笼子,营造了琳琅满目的集市情景。每个木箱上都绘有集市的场景片段。通道和连廊有货车的车轮印迹,端景是真实的货车,大面积的植物墙给整个空间带来了生机。二层的包间都有不同街景的再现。

艺术与生活不应是对立的,让艺术从画廊、博物馆走进大众生活,使艺术亲民化。本案设计以亲民的设计手法去表达,如老建筑结构、老家具、旧材料通过二次设计重新焕发新的生命力。这些元素有年代感、有故事、有回忆,体现出一种环保理念和传承的永续精神。

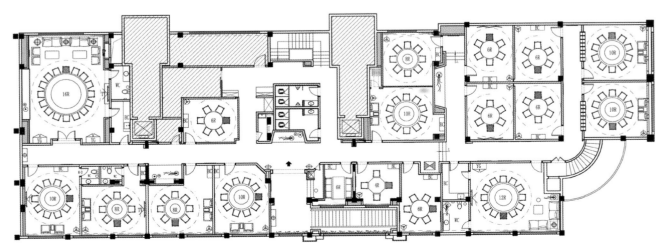

地点	面积	设计师	设计公司	主要材料
/广东深圳	/321 平方米	/王锟、刘进、叶俊峰	/深圳市艺鼎装饰设计有限公司	/旧木板、旧砖墙、花砖、钢板

Pepper Meets Chili
花椒遇见辣椒

In order to bring about the natural and unsophisticated beauty of the space that was formed during its long history, the designer used old bricks and tiles and painted screens, which reminder people of the old times, to create a dining environment with a lasting appeal and perfect structure.

As the restaurant targets the post-80s and post-90s, an "Encounter" theme was adopted in this case to combine tradition and fashion with Chinese elements as the basis. In the plane layout, an open design and convenient pass-through space would relieve traffic, while a romantic theme further enhances the temperament and taste of the space.

The width of the passageway satisfies people's needs of interacting with each other, but also keeps others at an agreeable distance. The shadow figures on the windows appear lively and vivid with perfect interaction between their colors and shapes and the light. The old wooden lamps hanging from the ceiling would attract people's attention and make them wonder.

为了让空间呈现出因历史文化而沉淀的古朴、自然的美感,设计师采用能渗透出时代气息的旧砖瓦及雕刻着丹青岁月的屏风,打造出富有韵味、结构完美的餐厅空间。

因定位消费人群为80后与90后,所以本案以"相遇"为主题,在兼顾了中式元素的同时,让传统和时尚碰撞出激烈火花。平面布局上,开阔的敞口设计和便捷的传菜通道,化解了餐厅的人流压力,加之唯美的主题渲染,使得空间气质和品位得到进一步升华。

过道的尺寸既满足了人们渴求交往、相遇的心理需求,同时也使人与人之间有一定的距离。皮影窗花的色彩、形态与光照相互呼应,惟妙惟肖。悬挂在顶棚的造型独特的老木吊灯,吸引了人们的眼球,让人想一探究竟。

地点	面积	设计师	设计公司	主要材料
/新疆乌鲁木齐	/1200 平方米	/蒋国兴	/叙品设计装饰工程有限公司	/壁布、深色实木地板、深色木饰面、黑木纹大理石、火山岩

Tori Talk
东篱·叙

The design inspiration of this project comes from the poem "Drink" of Tao Yuanming. The designer is not concerned with rigorous silence from traditional Chinese restaurants, but treats it as an artistic work. The designer applies traditional Chinese classical design elements with a new design technique, a perfect fusion of soft and warm dominant hue in the restaurant, offering people a sense of quiet conversion.

The designer takes account of privacy of the restaurant. The entrance leads people to a little ante-room, where there are classical wooden chairs by a door, original ecology of branch hanging drops on the wall, a round crock nearby, gurgling water making the quiet space more mysterious, several niches on the wall, warm lamplight casting its shadow on a coarse pottery, and a sliding door blocks people's vision, to make everyone have an infinite association to the scenery behind the door. The designer makes the passageway into a zigzag, and applies landscaped design technique, so there are different visual attractions in the every corner customers pass by, scenery everywhere, to a VIP box finally. It not only represents its beauty, but also retains its privacy and mystery.

Boxes in the restaurant are designed with a large number of original furniture and based on innovated furniture of Ming and Qing dynasties. Furniture is made by original ecology of woods which is not only environmentally friendly but also matching with the design concept of "back to nature, learn to live". This is what the designer and the owner would like to achieve. Boxes are full of vine wallpaper, which changes people's previous impression of boxes. Not only are boxes in a quiet atmosphere, but also a completely new sense of beauty and texture. Most of lamplights in boxes are soft circular arc, which means profound thought of traditional Chinese "round sky and square earth", to build a smooth feeling of spaciousness for the space, fused together with the overall warm style of the restaurant.

本案的设计灵感源于陶渊明的《饮酒》诗。设计师不拘泥于传统中式餐厅的严谨，而是把空间作为一个艺术品来对待，运用中国传统、古朴的设计元素，并加入新的设计手法，融合餐厅柔和、温暖的主色调，让人心里顿生一种皈依的宁静感。

设计师考虑到餐厅的私密性，在入口处设置了一个小前厅，门口放着木质的古朴的椅子，墙上是原生态的树枝挂饰，旁边是圆形的瓦缸、淙淙的水流，这一切让安静的空间显得更加神秘。墙上有几个壁龛，柔和的灯光投在粗犷的陶器上，一扇推拉门挡住了视线，让人对门后的风景产生无限的遐想。设计师把通道做成折线形状，运用了园林设计的手法，在顾客路过的每个转角都设有不同的视觉景点，步步皆景，最后到达一个VIP包间。这种设计既体现了美观，又保留了神秘感。

餐厅包间使用了大量的原创家具,设计以明清家具为参照,并做出了创新。家具采用原生态的木材,不仅环保,而且与这家餐厅的设计理念"回归自然、学会生活"相吻合。这正是设计师和业主想要达到的效果。包间多采用藤制壁纸,改变了以往人们对餐厅包间的印象,不仅使包间处于一种宁静的氛围中,而且创造了一种全新的美感和质感。包间灯具造型大多采用柔和的圆弧形,寓意中国传统文化中"天圆地方"的博大精深的思想,为空间营造一种流畅的通透感,与整个餐厅温馨的格调相吻合。

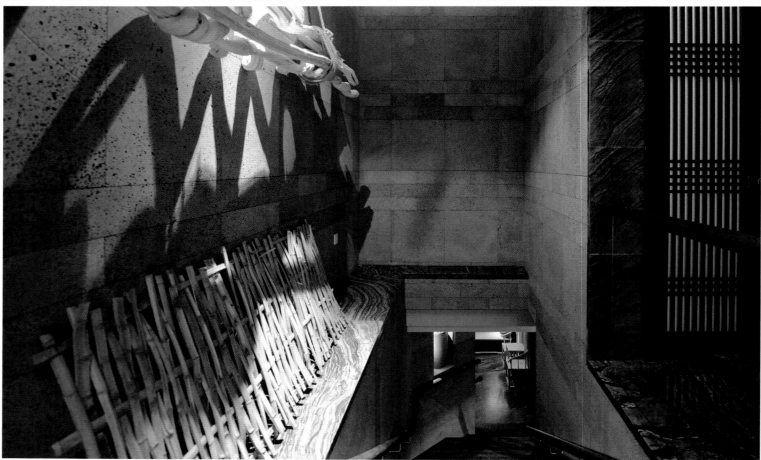

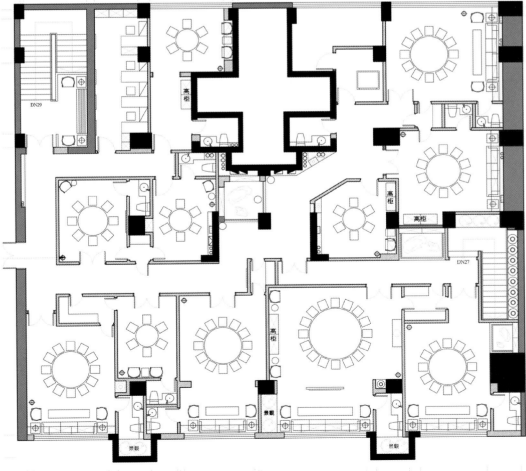

地点	面积	设计师	设计公司	主要材料
/山东潍坊	/130平方米	/汤善盛	/大石代设计咨询有限公司	/防火板、真石漆、方管、木地板砖

Maolu Impression Weifang Kaide Restaurant

茅庐印象潍坊凯德店

The project is a restaurant of an authentically flavoured Sichuan dishes. The owner would like to better express Sichuan culture by this design. The designer thinks authentic regional culture is actually a kind of living culture and a form that ordinary people accumulate and improve in a long term in daily life. Therefore, the designer brings nature to life, "to be life" and "to be scenery" are the guiding ideology of the project.

The project is inspired from special mountainous residence of Sichuan (especially in western Sichuan area), so at the door and the entrance of the project, several gables of residences with different heights leap to people's eyes, and there is a border full of plants and flowers at the bottom of the gables in order to be shared for the indoor scenery. The area between two gables forms an entrance of natural passageway, and several bamboo chairs and tables are scattered under the eaves, where people can have a rest and drink tea here(actually customers can wait for their dining seats of the restaurant). It adds a warm intimacy of ordinary daily life. The whole space is divided into several blocks, every of which is a "little house", to form a characteristic of houses in houses. Every block is relatively separate area, so as to increase privacy between blocks, and to faciitate owners to run their business just right.

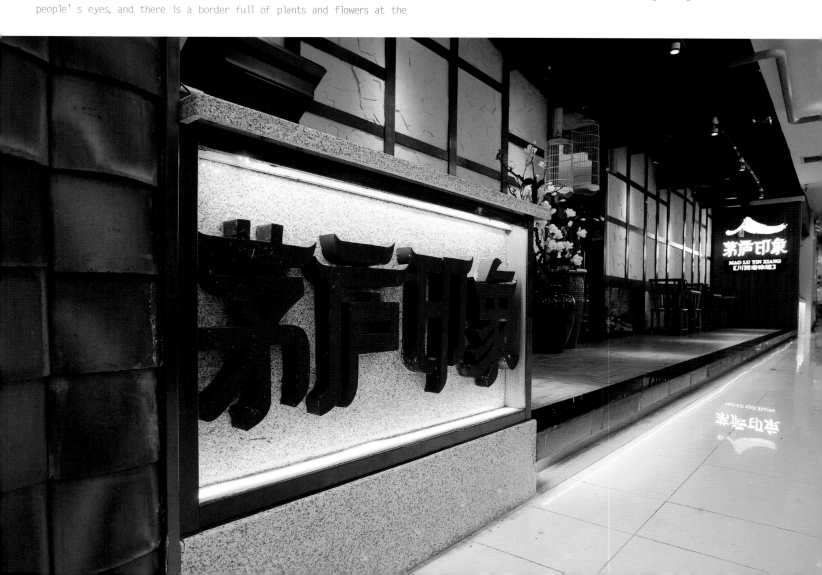

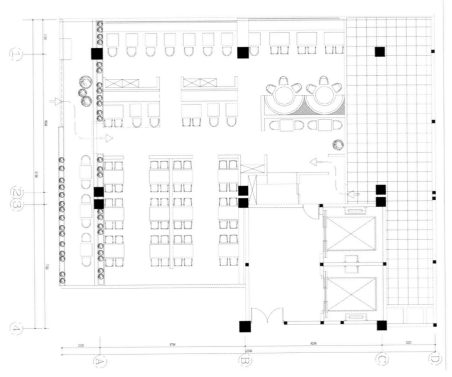

本案是一家做地道川菜的餐饮企业。客户想通过此次设计更好地表现四川文化。设计师认为真正的地域文化其实就是一种生活文化，是普通人日常生活中长期积累、沉淀并升华的一种形态。为此，设计师将自然引入生活，以"生活化""场景化"作为本案设计的指导思想。

本案设计的灵感来自四川特色的山地民居（特别是川西一带），为此，在门头与入口处设计了几座高低错落的民居山墙，山墙底部架空做成植物带，将景观引入室内，中间山墙与山墙之间预留了空隙，形成天然过道式的入口，而屋檐之下则散落几处竹椅茶桌，人们可在此休憩喝茶（实际是具有等位的功能），增加了市井生活的亲切感。

整个空间分成几块，每块都是一个"小房子"，形成了楼中楼的特征。每一块都是相对独立的一个小区域，这样既增加区域与区域之间的私密性，又便于客户适时地分区经营。

地点	面积	设计师	设计公司	主要材料
/山东潍坊	/130平方米	/汤善盛	/大石代设计咨询有限公司	/防火板、真石漆、方管、木地板砖

Maolu Impression WeiFang Wanda Restaurant

茅庐印象潍坊万达店

This case is a typical Sichuan restaurant offering authentic old western Sichuan taste for urban people. Bamboo and exposed brick on the first door design remind of mountainous western Sichuan residential characteristics, brought nature into life, making space "life" and "scene" and this is precisely the case design guidelines.

The interior design is to follow the "life" ideas and practices, bamboo erected beams, slightly propped panes and tiles stacked layers of facade make the indoor environment into outdoorsy, a bamboo chair in Sichuan also improved to a dining chair by designer, and Sichuan dialect word also hit the wall through the cast light, which increase the sense of space fashion. These distinctive regional symbols combined with each other, creating out this simple and stylish space.

本案是一家做地道川菜的餐厅，为都市人提供纯正的川西老味道。在门头的设计上，青竹、裸露的红砖等让人想到了具有川西特色的山地民居。

室内空间设计遵循"生活化"的思路与手法，竹竿搭起的房梁、稍稍撑起的窗格及灰瓦层层叠压的墙立面等都让室内环境室外化。四川的竹椅也经设计师改良成了餐椅，四川的方言也通过投字灯的方式打在墙上，增加了空间时尚感。这些鲜明的地域符号相互结合，造就了这个纯朴而时尚的空间。

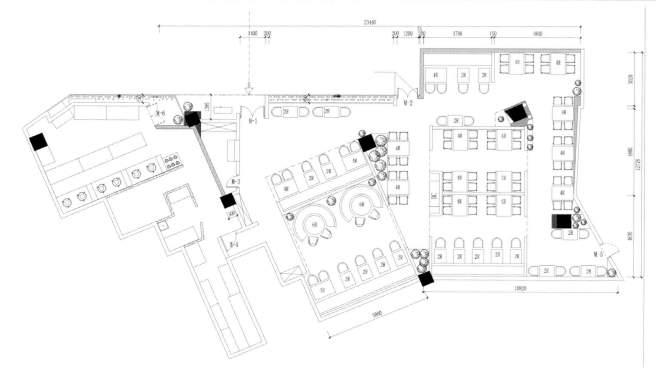

地点	面积	设计师	设计公司	主要材料
/新疆阿克苏	/1800平方米	/蒋国兴	/叙品设计装饰工程有限公司	/黑色花岗石、方钢、灰色砖、黑色砖、灰色壁纸、蓝色布艺

Mashijiu Pot Aksu Restaurant

马仕玖煲阿克苏店

The project is based on Chinese decoration, to build a fashionable, concise and modern Chinese restaurant by leaping color technique. The dominant hue of the space is grey, white and black, and blue as the decorative color enlivens the space atmosphere, which represents a sense of fashion.

The layout of the restaurant is clear with smooth moving line arrangement. The dominant hue of the ante-hall at the entrance is black. The designer skillfully applies lamplights in boxes through simple and smooth lines partition wall, very warm and sweet. The floor is paved by black granite like mirror, and reacts with bevel mirror at the top, which seems to extend the space, to make the ante-hall look broader. There are long Chinese chairs on the left side of the ante-hall in the waiting area, which humanizes to retain the function part to be harmonious with the surrounding environment.

When people first enter extra seats in the restaurant, what first leaps to the eyes is a circular booth like bird cage made by white round pipes. Curved lines soften hard Chinese concise lines. The semi-open bird cage partition doesn't block customer's sight, but a panoramic view of the whole hall. Customers can bypass the hall of extra seats to the boxes area of the restaurant. The designer takes the advantage of original space structure, and plans all sizes of boxes. The dominant hue of boxes is grey, with originally white-enameled ceiling, to keep the height of boxes. Well carved porcelain plate paintings on the wall, the concise Chinese chandelier and some exquisite decorations form an elegant, simple and beautiful dining space.

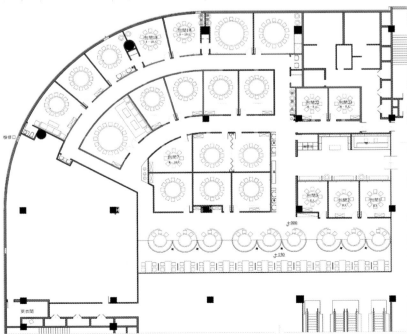

本案以中式风格为主，用跳跃的色彩打造一个时尚、简洁的现代中式餐厅。灰色、白色、黑色为空间的主色调，蓝色作为点缀色活跃了空间氛围，也彰显出时尚气息。

餐厅的布局明确，动线规划流畅。入口前厅以黑色为主色。设计师巧妙地运用包间的灯光，通过线条简单、流畅的隔墙，营造出温馨的空间。地面则用光洁如镜的黑色花岗石，其与顶部的斜边条镜上下呼应，拉伸了空间感，使前厅视觉倍感开阔。前厅左侧以长条中式凳为造型的等候区，也与周边环境十分融洽。

进入餐厅的散座区，映入眼帘的是以白色圆管组成的鸟笼圆形卡座。弧形的线条柔化了中式简洁线条的生硬感，半开放式的鸟笼隔断设计不会阻碍来客的视线，使整个大厅尽收眼底。绕过散座大厅就是餐厅包间区。设计师利用原空间的结构，规划了大小不一的包间。包间以灰色为基调，漆白的顶棚在新颖的同时也保持了原有的高度。墙面上精雕细琢的瓷盘艺术画、简约的主吊灯、精美的装饰品等形成格调高雅、造型简朴、优美的就餐空间。

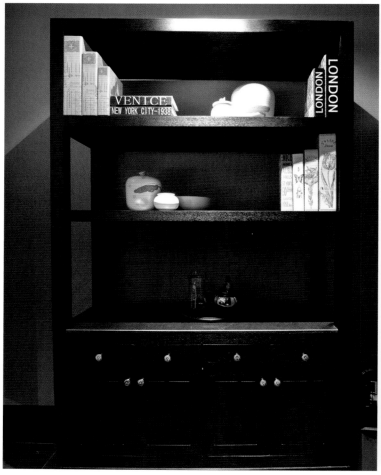

地点	面积	设计师	设计公司	主要材料
/新疆乌鲁木齐	/1200平方米	/蒋国兴	/叙品设计装饰工程有限公司	/黑色花岗石、稻草漆、木作蓝色做旧隔断、黑白配砖

Mashijiu Pot Changchun Road Restaurant

马仕玖煲长春路店

The project is another regular chain. The space style is modern Chinese style, whose classical Chinese elements are applied completely and vividly by modern technique. As for the application of colors, the designer continues to adopt black, white, grey and blue, and skillfully fuses them into the space, which makes the restaurant's atmosphere classical and warm.

The project design aims to be well-structured and well-proportioned. When people enter the restaurant, what first leaps to the eyes is the reception counter decorated by tracery, where it is infused with yellow lamplight, to be warm and beautiful. On the right side of the ante-hall, the labyrinthine corridor divides the extra seats area and boxes area on both sides neatly, and the circular boxes make the corridor become the beautiful scenery which cannot be ignored. On the left side of the ante-hall, there is the extra seats area of the restaurant after people go on several steps, and boxes are scattered on both sides. The classical tracery screen partitions extra seats orderly, so that every table of customers do not interfere with each other while dining, to build a harmonious dining environment. However, every piece of tracery is not exactly the same, but a patchwork of various styles of tracery. The boxes are concise and quaint. Quaint bricks, classical black and white paintings, Chinese traditional windows and Chinese tables and chairs react with the circular ceiling. Chinese lines separate the ceiling from the wall, not only to retain the original height of the ceiling, but also to partition the space effectively.

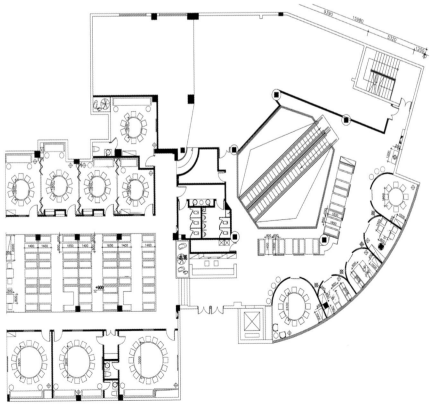

本案是马仕玖煲的另一家直营连锁店。空间风格为现代中式风格，设计师将古典的中国元素用现代的手法表达得淋漓尽致。在色彩运用上，延续了该品牌设计一直使用的黑、白、灰、蓝色调，通过将几种色调巧妙地结合，使餐厅的氛围古典而又温馨。本案设计的宗旨是层次分明、错落有致。进入餐厅，首先映入眼帘的是以花格装饰的吧台，散发着淡黄色的灯光使人感到温暖而美好。前厅右边，曲折的回廊两边整齐地排列着散座区和包间。圆形设计的包间让回廊成了一道不可忽略的风景线。前厅左边，踏上几个台阶便是餐厅的散座区，包间分散至两边。古典的花格屏风将散座区有序地隔开，使每桌客人用餐时都互不干扰，营造了一种和谐的用餐环境。然而每一块花格又不完全一样，各种样式的花格错落有致地交织在一块。包间的设计简洁而又不失古雅，古朴的小青砖、古典的黑白画、中国传统的窗格、中式的桌椅再配以圆形的吊顶与之呼应。中式线条将顶棚与墙面隔开，既保留了原有顶面的高度，又很好地划分了空间。

地点	面积	设计师	设计公司	主要材料
/四川成都	/2000平方米	/彭宇、许亮、王继、席庆、冯凯	/葵美树环境艺术设计有限公司	/黑木纹、绿玉、黄金洞石、爵士白、红洞石、樟木

South Pavilion Restaurant
南堂馆餐厅

In Sichuan province, in the late Qing Dynasty and early Republic of China, the project refers to the first-class delivery tavern. Now the theme element of the project is derived from delivery boxes. The designer adopts different materials, different colors as well as different scales to shape the theme structure of the space.

The designer sorts out the space by "one project, two characters, three lanes, four boxes room", which can represent different concepts of rich culture of Sichuan dishes, cooking technique, color culture and folk culture. Scenery changes by moving steps. Meanwhile, the designer applies the composition technique and lamplight consistent with contemporary aesthetic, to build a modest but noble feeing.

南堂馆在明末清初时的四川地区泛指高级的外送酒肆。本案以送酒席时的抬盒为主题元素，运用不同材质、不同色彩及不同尺度的抬盒来配合空间的主题造型。

整体以"一间南堂馆、两种性格、三条巷子、四种包房"来梳理空间，分别形成了表现川菜的文化底蕴、烹饪技法、色彩文化、民俗文化的不同概念的区域，使顾客能感受到步移景异。同时，设计师运用符合当下审美的构成手法、灯光形式来烘托餐厅设计带来的高贵感受。

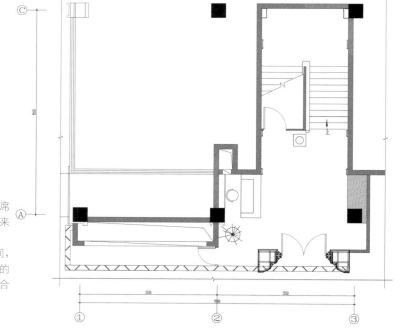

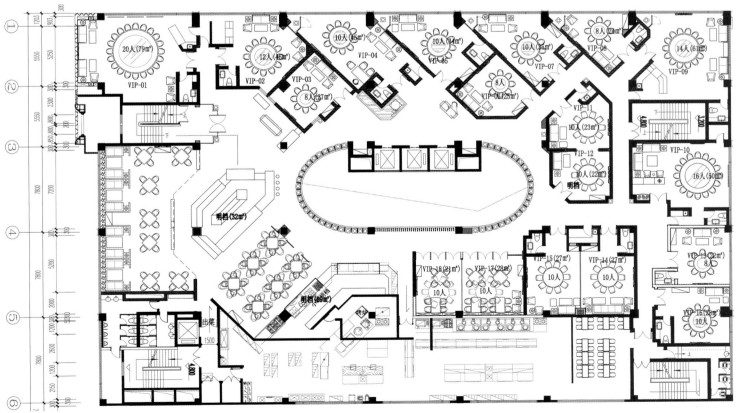

东方韵律 新中式餐饮空间设计

地点	面积	设计师	参与设计	设计公司
/福建福州	/1700平方米	/林鸿	/李加平、蔡钢渤、魏子灏	/维野商业空间设计

Spicy Talking Sichuan Restaurant Daming Outlet

说麻道辣川菜馆达明店

Brick, wood and glass intersect with each other in the door head, while two quaint stone sculptures stand in both sides of the door, reminding people of an old inn in ancient China. Coarse wall texture and grey bricks can be seen through the glass, making people curious to open the door and find out what's inside.

Inside the restaurant, a Chinese-style design has created an atmosphere totally different from that of traditional Sichuan restaurants, showing a sense of calmness and composure described in the ancient Chinese poet Tao Yuanming's famous line "My eyes fall leisurely on the Southern Mountain". The atmosphere is a little stagnant but not depressing, with large areas of bright red offsetting the reservedness of grey bricks and brown wood. Just like the indispensable chili pepper in Sichuan cuisine, the red color is the undertone of Chinese people's feelings, which with its straightforward hospitality gives you a good appetite. Eating has become a style which brings serenity of the heart and the mind.

The designer has ingeniously integrated Sichuan culture into the space: the traditional facial masks scattered around in the room and mahjong tiles on the walls are definitely the most representative elements of Sichuan, the "Land of Abundance". The box design in this case has broken the traditional private design, but has made the space more transparent and dynamic with wooden grids. Tables, chairs and decorations have the simplest texture and shape, demonstrating elegance and great taste.

门头设计中，砖木和玻璃交错而至，古朴的石雕分立门边，颇有旧时客栈的感觉。透过玻璃，隐约可见粗砺的墙面肌理、青灰的墙砖，让人不由心生好奇，想要推门而入、一探究竟。室内空间中，中式风格的设计打造出了与传统川菜截然不同的就餐氛围，有一种"悠然见南山"的从容与淡定。餐厅的气氛虽沉却不闷，跳跃的红色在空间里大面积使用，缓和了灰砖、棕木的内敛，一如川菜里那不可或缺的火红的辣椒，更是让人感受到有一种热情直入心间，令人食欲大增。吃，成了一种格调，实现了心与境的完美融合。

设计师巧妙地将川蜀文化融入空间中，散落在空间里的脸谱、悬于墙面上的麻将，无疑是天府之国极具代表性的元素。在隔断的设计上，本案空间则完全打破了常规的私密设计，木格栅若隐若现，让空间更显通透的同时，增加了空间的视觉动感。桌椅、摆件的材质与造型力求简洁、大气而富有质感。

地点	面积	设计师	软装设计	设计公司	主要材料
/福建福州	/380平方米	/李川道、郑新峰	/陈立惠、张海萍	/福建东道建筑装饰设计有限公司	/手工陶砖、特纹玻璃、复古木板、肌理漆、仿古地砖

Chating Restaurant Qiaoting Fish Town
桥亭活鱼小镇茶亭店

What does the project present is an innovative new-Chinese-style restaurant, not only being straightforward and uninhibited like ancient inns, but also being elegant like western restaurants. Various elements merge into the space to form a distinctive space.

Concise and neat integrated lines, casual layout and semi-transparent space partition not only meet the requirements of aesthetic, but also avoid the sense of oppression, to realize space requirements of being open and practical. Not carved log materials support and partition the inner space, building a natural and simple space atmosphere. Traditional Chinese space always offers people a feeling of elegance, added with some fresh colors, and makes the space expression richer, where both the pastel wall of cool color and colorful striped decoration at any place enliven indoor atmosphere just right. It is worth mentioning that indoor scenery creation techniques and carefully selected decorations through skillful collocation and combination not only highlight the restaurant style, but also spread a quiet and distant oriental verve inside the space. Unique decorative paintings, elegant panes, exquisite lamps and lanterns and various kinds of ornaments successfully create a dining space with lingering charm of the Orient.

In spare time, people can invite their friends to have a drink of some cups, or enjoy good food to imitate the elegance from ancient people, to seek for a quiet in this bustling city. Maybe, this is the charm of Chinese style space.

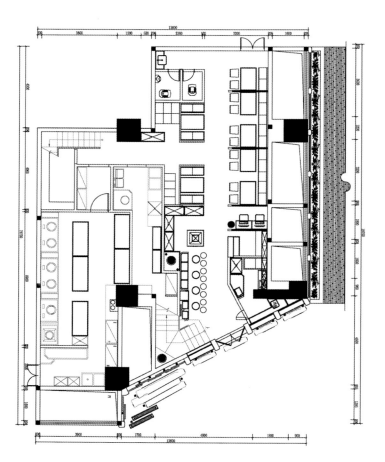
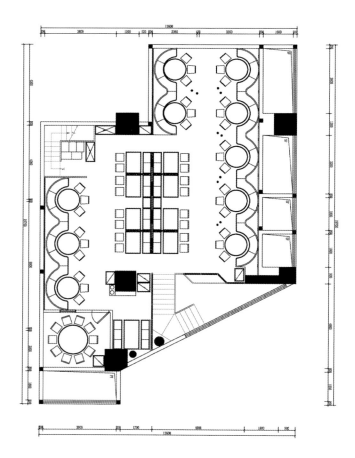

本案展现的是一个别具新意的新中式风格餐厅，既有古代客栈的粗犷，也有西方餐厅的优雅，结合各种元素，混搭出与众不同的空间。

简洁利落的整体线条、随意的布局、半通透的空间分割等不仅满足了美观的要求，还巧妙地避免了压抑感，满足了开敞、实用的空间需求。未经雕琢的原木材质在室内起着隔断与支撑的作用，营造出淳朴自然的空间氛围。传统中式空间向来给人以素雅之感，随着一些鲜活色彩的加入，空间的表情变得更加丰富。无论是冷色系的粉彩墙面，还是随处可见的彩色条纹装饰，都恰到好处地调动了室内气氛。值得一提的是，室内的造景手法、精心挑选的装饰物经过巧妙地搭配与组合，既突显出餐厅的格调，又让清幽淡远的东方神韵在室内蔓延开来。独特的装饰画、清雅的窗格、别致的灯具及各种考究的摆件，成功打造出具有东方灵性的就餐空间。

闲暇时，邀约好友来此小酌几杯，或品尝美食，附庸一下古人的风雅，于熙熙攘攘的都市之中，寻找一片宁静。也许，这正是中式空间的魅力所在。

地点	面积	设计师	设计公司	主要材料
/山东济南	/400平方米	/王远超	/思锐空间设计有限公司	/灰色大理石、铁刀木、稻草漆、镜面、钨钢、清玻

Hui Jiangnan Private Kitchen

汇江南私房菜

By extracting the Southern traditional architectural, and the perfect combination of modern and tradition, designers use modern, contracted decorative techniques and materials application, reveal material environment, situation, mood, as far as possible to perform the euphemistic and ethereal of Gangnam.

In space design, designer use gray stone and chestnut wood in ceiling and wall trim, matched with white walls and gray tiles, extremely fresh and graceful. While the southern elements skillfully blend together, oiled paper umbrella combined with chandeliers in the central nave, depicted verse about Jiangnan by the ancients om wall and ceiling, swimming fish in the pool, brush of the poet, known as "silent poetry" of Taihu stone, Jiangnan Yan in the mirror, peach blossom drawing from master, pathetic radian canna, all makes people feel the space full of poetic. Water flowing down the sink along the pool, when enter in, guests can hear gurgling water but do not know where the water comes from, until walking near, a pool of fish just come into view, observing the fish swim, forget the vulgarity.

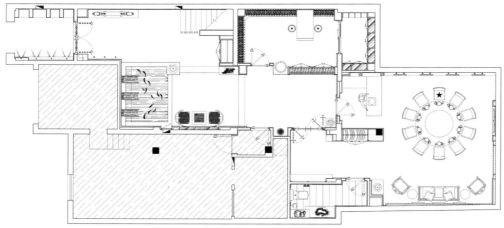
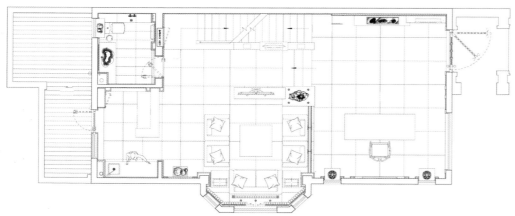

设计师通过对江南传统建筑的提炼，使现代和传统完美结合。用现代、简约的装饰手法及材料展现物境、情境、意境，尽可能地表现出江南的委婉、灵动。

空间设计中，设计师用灰色大理石与铁刀木做顶棚和墙体边饰，与白墙灰瓦相配，尽显清新雅洁。设计师把江南元素巧妙融会贯通于其中，中厅的油纸伞组合吊灯、墙壁及顶棚上的古人描绘江南的诗句、池中的游鱼、墨客的毛笔、被誉为"无声的诗"的太湖石、镜中的江南燕、大师的桃花图、那有着令人怜惜弧度的美人蕉，都让人感觉到空间里充满着诗情画意。池中水顺着水槽往下流淌，宾客进门便可听见潺潺水声却不知水在何方，直至走近了，一池游鱼方才映入眼帘，静观此景令人忘俗。

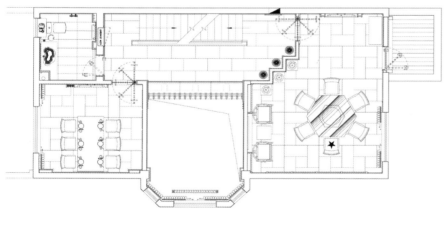
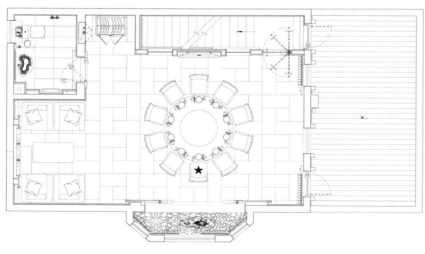

东方韵律 新中式餐饮空间设计

地点	面积	设计师	设计公司	主要材料
/四川成都	/2000 平方米	/杨凯、徐颖聪	/黑蚁空间设计工程有限公司	/木地板、地砖、壁纸

Rongfu Restaurant
蓉府餐厅

The project is defined to high quality customers, but different from traditional luxury of restaurants. The project is totally modest in space, not aggressive, with details to emphasize the artistic temperament of the space.

The space is designed according to modern Chinese style and combined with a large number of original arts in details, which easily leaps to people's eyes, and attracts a large number of people with high culture taste. Here the unique cultural atmosphere consistent with high quality space is the major reason to attract people.

The space is decorated with logs in a large area to convey a sense of natural luxury. At the same time, gradations of the space is expressed by a contrast way, such as: the texutre details contrast of polish and a gloss finish by the same stone materias in the same place. The boxes are designed by flexible and changeable techniques, more suitable for the demand of diversed functions.

本案定位于高端市场，但有别于传统豪华的高端餐饮空间。本案空间整体低调、有细节且着重突出空间的艺术气息。

空间采用了现代中式风格进行设计，在细节处融入了大量的原创艺术，让人眼前一亮，获得了大批文化品位较高人群的青睐。这里独有的文化氛围和符合顾客身份的高档的空间感受是吸引他们的主要原因。

空间大面积使用原木装饰，表达出一种自然、奢华的感觉。同时采用对比方式来表现空间的层次，如同样的石材在同样的地方表达出光面和烧面的质感细节对比。

包间采用灵活多变的手法设计，更加适合多样化的功能需求。

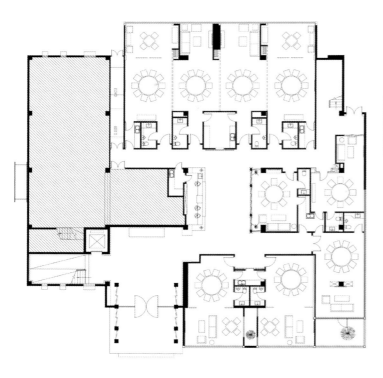

地点	面积	设计师
/浙江宁波	/400平方米	/卢忆

Sanshili Alley Restaurant
三市里胡同餐厅

The original design plan is to hope customers enable to enjoy the pleasant dining environment, which resonates with the restaurant and the entire space design. The instrument as a design element is applied throughout the whole space, and the major material of the restaurant is bamboo, to match with the elegant old architectural structure of the restaurant.

At the entrance of the restaurant, two lounges, tables and chairs made like drums and bamboos scattered on the wall represent music rhythm and react with the theme. The lamp above the cashier counter is applied with the flute hidden in the ceiling to be a natural lighting source. Allegros are connected to be a wall for the partition on the second floor, which breaks the application of traditional partition and can have a vague vision behind the partition through gaps between allegros. The ceiling of the restaurant's boxes is decorated with concise style with the bird cage as the common items decoration to make customers enjoy themselves in a certain extent. Bamboo allegros between boxes are lined and arranged to form the pattern to build a folding door, so as to meet the customers' demand of custom-made sizes of boxes.

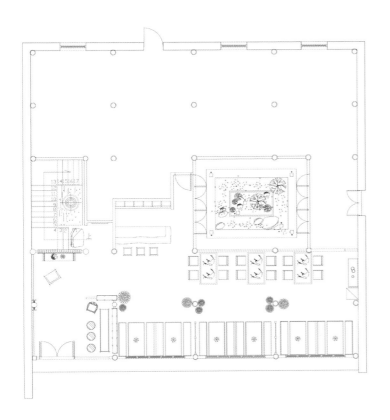
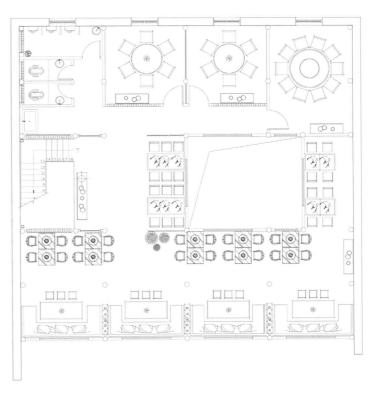

本案最初的设计理念是希望顾客能享受愉悦的用餐环境，与餐厅及整个空间设计产生共鸣！"乐"同"乐"，所以运用了乐器作为设计元素贯穿整个空间。餐厅的主要材料为竹材，其与餐厅的老建筑结构的风雅感十分协调。

餐厅入口处设置了两个休息等候区，以乐器鼓做成桌、凳，竹节高低错落地镶嵌在墙体上，示意音乐韵律，呼应主题，收银台顶部照明运用了笛子暗藏灯带，做成天然照明源。二楼隔断用快板串联成墙，打破对隔断的传统设计，透过快板的空隙可以隐约领略隔断后面的风情。餐厅包间顶面装饰采用简约的风格，用鸟笼这种常见的物品作为装饰，使空间极富趣味性。包间与包间之间用竹筒切片纯手工排列成图形制作成折叠门，满足顾客对包间大小的自定义需求。

地点	面积	设计师	设计公司	主要材料
/四川成都	/300平方米	/高雄、高宪铭、郭予书	/道和设计机构	/乳化玻璃、烤漆玻璃、玫瑰金、橡木饰面板、夹膜玻璃、雅士白大理石

Shicai Yunnan Cuisine Restaurant

食彩云南料理餐厅

"Through the zigzag path to a Buddhist temple surrounded by trees and flowers, birds sing and dance in the mountain, and a lake purifies people's heart."

The design thought is derived from such dreamland. The layout of the zigzag path uncovers delicate and exquisite gradations. Fresh oak color materials in the space convey a harmonious indoor atmosphere. A large area of glass is applied in the space, which seems to enlarge the space as well as reflect the space like a clean and clear lake. The pure peacock-blue glass just like the sky is carefully reacted with flowers, the grass and trees, full and not losing rhythm. Colorful butterflies fly in the sky, with strong vitality. The fusion of nature and dynamics seems to be able to breathe, vital and vigorous.

It is matched with folk specialty of Yunnan Province, the unique totem and the distinctive geographical landscape image. The delicate combination of nature and nationality quality fusing to the design is a real definition of the space.

"曲径通幽处,禅房花木深。山光悦鸟性,潭影空人心。"
本案设计以这样一种意境为参照。曲径通幽的布局,揭开层层精致与细腻。空间中清新的橡木原色饰面板营造着和谐的室内氛围。空间中使用大面积的玻璃设计,在视觉上扩大空间的同时让空间如湖水般清澈。有如天空般纯净的孔雀蓝玻璃与鲜艳的花卉草木细心镶嵌,饱满而不失节奏。缤纷的蝴蝶在空间中交错,带着勃勃生机。自然与动态的融合,好似会呼吸般生机盎然。
整个空间呈现出云南的民族特色,拥有独有的图案与鲜明的地域景观画面。将自然与民族特色融合是本案设计的最精彩之处。

地点	面积	设计师	设计公司
/天津	/600平方米	/熊华阳	/深圳市华空间设计顾问有限公司

Qian Yuan Fashion of Tianjin
天津乾园风尚

This case is a particularly Jiang-Nan style catering space. The stylist emphasizing on the poetic beauty, Chinese ink painting and special garden art integrate the nature with perfection subtly. To reveal the tenderness and scholarliness of Jiang-Nan naturally, the stylist crafty absorbs these element from the hollow doors and windows in Jiang-Nan garden, the Chinese ink painted lotus mural, the special suspended lamp, freehand brushwork in traditional Chinese paintings, Chinese style screen and so on, combining with the leather that gives us the contemporary feeling extremely, and naturally build give a kind of fashion but also concise and elegant dining atmosphere. In space, stylist subtly divide restaurant into different dining area, expressing its artistic conception.

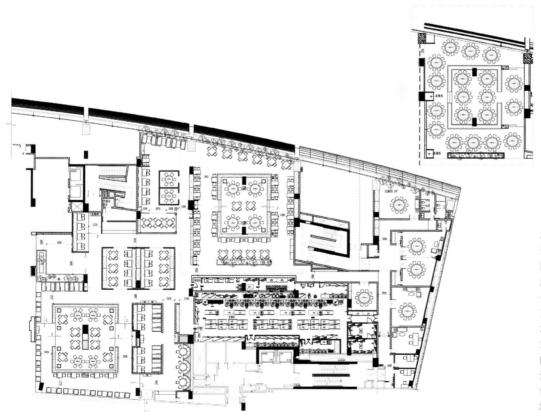

本案是一家极具江南风格的餐饮空间。设计师将江南水乡的诗情画意、水墨丹青、园林艺术等特色，自然、完美地融合在一起。为了自然地展现江南的柔情、细腻、书卷气，设计师以大理石、实木、白砖为主要材料，巧妙运用江南园林的镂空门窗、水墨荷花壁画、特色挂灯、写意国画、中式屏风等中式装饰元素，结合极具现代感的皮革，顺理成章地营造出一种时尚、简洁、雅致、宁静的用餐氛围。在空间布局上，设计师巧妙地将餐厅分隔成不同的用餐区域，各具意境。

地点	面积	设计师	设计公司	主要材料
/四川成都	/300 平方米	/高雄、高宪铭	/道和设计机构	/玻璃钢雕塑、波浪板、夹膜玻璃

Taste Talk Restaurant
味语餐厅

Chengdu has a long history, profound culture, it is territory of Shu in the ancient, and the city was built in the Qin by combined with Ba and Shu. Its specialty is bamboo, endowed them with beauty and meaning.

This case combined the lines which represent modern Western style with the block surface, forming a space with overall rhythm, plus with the unique tone of bamboo from the East, in the hollow carved Craftsmanship, making people relaxed and happy. Deep black marble table and pure gray, white leather sofas contrast, resulting in a unique contrast beauty, enable fashion and ancient atmosphere blended perfectly.

The overall style of space is modern Chinese style, in the light of contrast, ancient ink bamboo wall paintings make the restaurant manifests the ancient atmosphere, the state of quiet beauty seem ready to come out.

地点	面积			主要材料
/四川成都	/300 平方米			

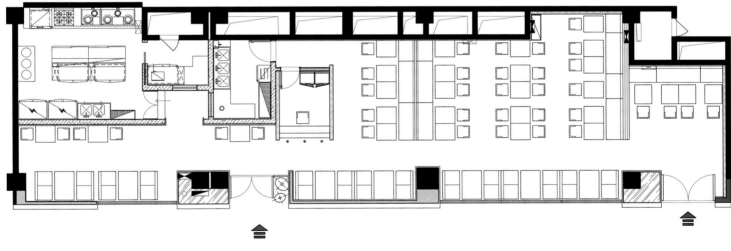

成都历史悠久，文化底蕴深厚，古为蜀国地，以竹为特色，以之为美，赋之以意。本案将具有新中式风格的线条与块面相结合，使空间彰显出整体感和韵律感。同时结合竹子特有的形式，在镂空精雕的工艺下，让人心旷神怡。大理石桌面深邃的黑色与皮革沙发纯粹的灰、白色形成对比，产生出独特的对比美，让时尚与古老的气息完美融合。

空间整体风格以现代中式为主，古墨色苍竹墙绘在灯光的烘托下，让餐厅彰显出古韵气息，静美之态呼之欲出。

地点	面积	设计师	设计公司	主要材料
/福建福州	/335 平方米	/李川道、郑新峰、陈立惠、张海萍	/福建东道建筑装饰设计有限公司	/手工陶砖、玻璃、复古木板、肌理漆、仿古地砖

Story of Little Town Retro Romantic Feelings

小城故事复古情怀

This case use reminiscent tone to set up a corridor of time, making people feel like through the veil time and return to simple years. Before enter to the space, you have go through a long corridor, where gray old furniture on the floral collage tiled floor, specially built roof with grey tiles, and a wall full of old photographs, indicating we will enter a little different world.

In order to concert with Qiaoting brand, store design revolve around ancient town elements, especially built a whole space as the appearance of town, a variety of roof tiles, wooden windows, brick walls as well as the stone bridge, giving immersive realism. Unadorned concrete floor just has the most simple polished management, store and chairs are the most common form of the deck of the 1980s, partition between the seats is a wrought iron frame, additional stained glass and barbed wire, the air full of rust old atmosphere.

Abandoned wooden boat, old window frame has revived here, they are re-spliced and combined, paint a different color, they form a barrier wall to replace the boring rigid white walls, so that each area are connected, natural the formation of a small box, through different forms of windows, everywhere is a landscape.

本案以怀旧为基调设计走廊，人们仿佛穿越了时空又回到那质朴的岁月。进入空间之前要经过长长的回廊，这里有碎花拼贴的瓷砖地面、采用了灰白的做旧家具、特意打造的青瓦的屋檐和满满一墙的老照片，预示着我们将进入一个独特的小天地。

为了呼应桥亭品牌，店内设计围绕古镇元素，特将空间整体打造为古镇模样。各种砖瓦的屋檐、木质的窗户、青砖的墙面还有石桥，给人身临其境的真实感。

不加修饰的水泥地面，只做了最简单的磨平处理，店内的桌椅均采用20世纪80年代最常见的卡座形式。座位间的隔断是铁艺的门框，再搭配彩色玻璃与铁丝网，使空气中充满着铁锈的老旧气息。

废弃的船木、老旧的窗框在这里得到了新生。它们被重新拼接组合，粉刷上不同的色彩。它们形成了墙面隔断，替换了枯燥死板的白墙，让各区域间形成连通，自然地形成小包厢。透过形态各异的窗子望去，处处都是风景。

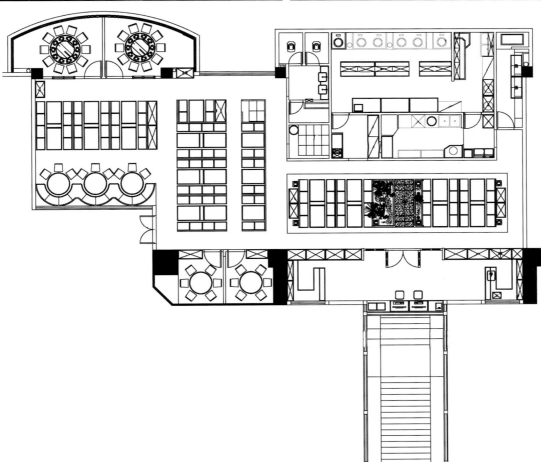

地点	面积	设计师	设计公司	主要材料
/新疆乌鲁木齐	/1200 平方米	/蒋国兴	/叙品设计装饰工程有限公司	/灰色仿古砖、白色硅藻泥、深色实木地板、浅色地砖

Number.1 Zhuxi
竹溪一号

The case is located in Urumqi Nanhu road of Xinjiang, where shuttled in the crowd with unique geographical location, is a specialty restaurant main in Chinese food.

In this case, designer combined the traditional rustic style with modern Chinese style elements, creating an atmosphere with fresh, cheerful to merge together with melody, so that space is not dull. As to the color, designer use dark green as main colors of the restaurant, white, yellow as interspersion, slightly with retro furnishings, matched with soft lighting make the space very cheerful and fresh.

Aisle and rooms are divided by septal fenestra, and big rooms use folding screen to divide space making room for rational use. The combination of dark green false beams and white oak branch splice used in the ceiling, has promoted the taste and beauty of space. Dead wood, pebbles, bamboo, fabric and other natural materials has been used as soft decoration in aisle and hall; The decoration of green plants, porcelain, large pots and other furnishings, has assimilated the traditional decoration features as "shape", "spirit", making the restaurant more culturally charm and artistic conception, reflecting the unique charm of Chinese traditional food culture.

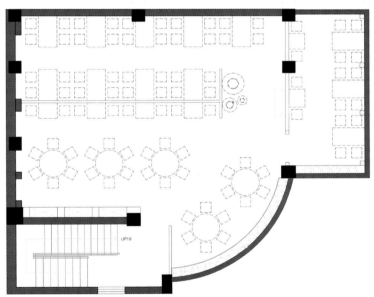

本案位于新疆乌鲁木齐南湖路，人流穿梭不息，地理位置得天独厚，是一家主营中餐的特色餐厅。

本案中，设计师将传统的田园风格与现代中式风格元素相结合，以营造清新、欢快的氛围为主，令空间不呆板。而色彩方面，设计师以墨绿色为餐厅的主要色调，白色、黄色作为点缀。略带复古的家具，再配上柔和的灯光，使空间十分欢快而又清新。

其中过道、包间多采用大的隔窗来分隔空间，大包间采用了屏风隔断划分空间，使空间得到合理利用。顶棚采用墨绿色的假梁和白色橡木拼条，提升了空间的趣味性和美感。大厅、过道在装饰上均使用枯木、鹅卵石、竹、织物等天然材料。绿色盆栽、瓷器、大陶罐等摆设，吸取传统装饰"形""神"的特征，让餐厅更具有文化韵味和意境，体现出中国传统餐饮文化的独特魅力。

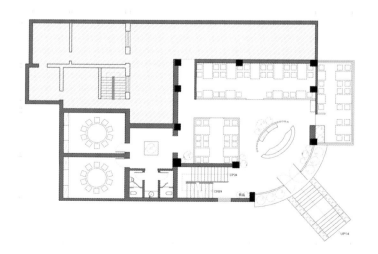
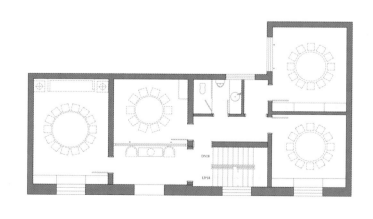

地点	面积	设计师	设计公司	主要材料
/河北石家庄	/700平方米	/张迎军	/大石代设计咨询有限公司	/原木、涂料、铁板、卵石

The Mutton Soup of Eight Banners

八旗羊汤

The soup is produced from the Mulan Paddock with high quality ingredients and their traditional simple and original cooking flavor. The project is planned from original life experience on the grassland. Single materials, concise lines and exquisite details divide the space into many parts of areas according to the functions and requirements.

The log (birch) furniture of grassland impression, a variety of furniture styles, refined materials of earth, wood and stone with modern technology, the materials of iron, wood and stone to retain time and DIY items made from twigs and pebbles build the casual and enjoyable dining space.

Melodious ditty of huqin, the sketch of lovely animals, insects, mushrooms and golden lotus on the grassland as well as the poetry and essays from the grassland bring a fresh original life experience to the urban people, who would produce the resonant imagery of cool and pure earth, temporarily forget the tense work and complicated life trivia, and at this moment enjoy happy time and original delicious flavor —original, natural, free and easy.

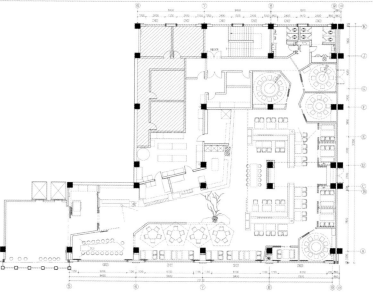

八旗羊汤的定位是用来自木兰围场的优质食材以八旗人传统的简单原味烹饪方法加工。而本案的设计定位，则来自对坝上草原的原味生活体验。单一的材质，简约的线条、精致的细节等将空间按需求和功能分成若干单元区。

来自草原印象的原木（桦木）家具，多样化的家具形式，现代工艺的土木石的粗材细作，留住岁月时光的铁木石质地，与树枝、卵石制作的DIY小品，共同打造出这个随性、写意的餐饮空间。

悠扬的胡琴小调、可爱的草原小动物、昆虫及金莲花形象速写，还有来自草原的诗歌美文，带给城市人清新的原味生活体验，共鸣梦想中的清凉净土，暂忘紧张的工作和繁杂的生活琐事，享受这片刻的幸福时光和原汁美味，体会原味、自然、自由、自在。